A MASTER REGULATOR OF OXIDATIVE STRESS - THE TRANSCRIPTION FACTOR NRF2

Edited by **José Antonio Morales-González, Ángel Morales-González** and **Eduardo Osiris Madrigal-Santillán**

A Master Regulator of Oxidative Stress - The Transcription Factor Nrf2
http://dx.doi.org/10.5772/62743
Edited by José Antonio Morales-González, Ángel Morales-González and Eduardo Osiris Madrigal-Santillán

Contributors

Angela Adamski Da Silva Reis, Rodrigo Da Silva Santos, Aline Helena Da Silva Cruz, Elisangela Gomes Da Silva, Aparecido Divino Da Cruz, Gustavo Rodrigues Pedrino, Lewis Cantley, Gina DeNicola, Ting-Yu Lin, Gabriele Siciliano, Annalisa Lo Gerfo, Tomas Alejandro Fregoso Aguilar, Brenda Carolina Hernandez Navarro, Jorge Alberto Mendoza Perez, George Mellick, Michael Todorovic, Ola Hammarsten, Sherin Mathew, Claudia Camelia Calzada, Carlos Alberto Jiménez Zamarripa, Esther Ocharan, Gisela Gutiérrez Iglesias, Hsiu-Ni Kung, Jhih-Pu Syu, Jen-Tsan Chi, Rosângela Araújo, Danyelly Martins, Maria Amélia Borba, Peter Zolotukhin, Anna Belanova

Published by InTech

Janeza Trdine 9, 51000 Rijeka, Croatia

Publishing Process Manager Romina Rovan

Technical Editor SPi Global

Cover InTech Design team

First published December, 2016

Printed in Croatia

Additional hard copies can be obtained from orders@intechopen.com

A Master Regulator of Oxidative Stress - The Transcription Factor Nrf2, Edited by José Antonio Morales-González, Ángel Morales-González and Eduardo Osiris Madrigal-Santillán

p. cm.

Print ISBN 978-953-51-2837-3
Online ISBN 978-953-51-2838-0

For EU product safety concerns:
IN TECH d.o.o., Prolaz Marije Krucifikse Kozulić 3, 51000 Rijeka, Croatia,
info@intechopen.com or visit our website at intechopen.com.

Meet the editors

José Antonio Morales-González engaged in doctoral studies in Biological Sciences (with a specialty in cellular biology) at the UNAM. Dr. Morales-Gonzalez has been awarded with diverse recognitions: the Alfonso Caso medal for academic merit by the UNAM and distinguished by the National System of Researchers (SNI) as National Researcher level 1 (2010–2016) and level 2 (2017–2020). He is the author of 40 internationally published articles and editor and coordinator of 27 specialized books. Dr. Morales-González is a full-time Titular Professor-Researcher at the ESM-Instituto Politécnico Nacional, México.

Ángel Morales-González engaged in doctoral studies specializing in Systems Engineering at the SEPI-ESIME, IPN. He was distinguished by the National System of Researchers (SNI) as a National Researcher candidate level (2016–2018). He is the author of 13 internationally and nationally published articles and editor and coordinator of 5 specialized books. In this moment, Dr. Morales-González is a full-time Titular Professor-Researcher C at the ESCOM-Instituto Politécnico Nacional, México.

Eduardo Osiris Madrigal-Santillán completed his PhD in Chemistry and Biological Sciences with the Genetic Toxicology specialty from the National Polytechnic Institute. Dr. Madrigal-Santillán has been recognized as a National Researcher level 1 awarded by the National System of Researchers belonging to CONACYT-Mexico (2010–2016) and level 2 (2017–2020). He is the author of 40 internationally and nationally published articles and editor and coordinator of 13 specialized books. Dr. Madrigal-Santillán is a full-time Titular Professor-Researcher at the ESM-Instituto Politécnico Nacional, México.

Contents

Preface

In recent times, attention has been drawn to the participation of transcriptional factor Nrf-2 as a modulator of diverse cellular functions, especially cytoprotection.

Among the functions reported for factor Nrf-2, we find that it regulates the antioxidant defenses, participates in cellular proliferation, modulates antiapoptotic as well as proapoptotic processes, and helps in cellular detoxification, among others. The first part of this book is focused on discovering the participation of Nrf-2 in cytoprotection. It begins with the review of the endogenous antioxidants and their role as protectors of the cellular functions. In addition, in this section, diverse chapters delve deeply into the role that Nrf-2 plays in the regulation of the metabolism, of oxidative stress, and of the inflammatory process and the participation of phytochemical agents in oxidative stress and in the activation of Nrf-2. Likewise, it approaches the regulation existing between Nrf-2 circuits with AP-1 and, on the other hand, also the role of KEAP1.

It has been found that Nrf-2 participates in the prevention of diverse diseases. Similarly, once the Nrf-2 disease is established, it continues to possess a certain participation. While there is a good side of Nrf-2 as a cytoprotector agent, Nrf-2 also has a "bad" side: for example, it has been reported that it participates in the resistance that certain cancerogenous cells have to chemotherapy. In the second section of the book, the authors approach themes with respect to the participation of Nrf-2 in diverse diseases, such as diabetes, neurodegenerative diseases, and Parkinson disease, and, in general, on the participation of oxidative stress in diverse diseases.

Due to that at present, the majority of diseases are associated with alterations in oxidative stress and inflammatory processes, and in that Nrf-2 is a modulator of these processes; knowing how this transcriptional factor functions and is regulated opens a therapeutic window to diverse diseases. Therefore, the efforts of various investigation groups are centered on finding activators and/or inhibitors of Nrf-2 to prevent or control diverse diseases, for example, cancer, where it would be important to regulate Nrf-2 in order for it to activate apoptosis pathways in cancerogenous cells, or in neurodegenerative diseases where cell death is predominant, it would be important for Nrf-2 to activate antiapoptotic pathways.

Ever-increasing investigation is being published on the role that Nrf-2 plays in cellular functioning and in diverse diseases, with surprising results in terms of how complex the modulation of Nrf-2 is. As Michalopoulos notes in reference to Nrf-2 and liver regeneration: "Unexpected and contradictory signaling functions are often discovered, which in retrospect are better understood as providing a fine balance for optimization of processes which the cell understands better than we do. Nrf-2, as an umbrella protector for the hepatocytes, exercises a very important function in perhaps slowing things down, when needed, to prevent complete catastrophe. This may be one of its most important functions."

Dr. José Antonio Morales-González
ESM-Instituto Politécnico Nacional, México

Dr. Ángel Morales-González
ESCOM-Instituto Politécnico Nacional, México

Dr. Eduardo Osiris Madrigal-Santillán
ESM-Instituto Politécnico Nacional, México

Section 1

Nrf2 a Modulator of Cytoprotection

Chapter 1

Endogenous Antioxidants: A Review of their Role in Oxidative Stress

Tomás Alejandro Fregoso Aguilar,
Brenda Carolina Hernández Navarro and
Jorge Alberto Mendoza Pérez

Additional information is available at the end of the chapter

http://dx.doi.org/10.5772/65715

Abstract

Oxidative stress (OxS) constitutes a disturbance caused by an imbalance between the generation of free radicals and antioxidant system, which causes damage to biomolecules. This, in turn, may lead the body to the occurrence of many chronic degenerative diseases. Therefore, it is very important to know the functioning of those endogenous (and exogenous) antioxidants systems to prevent such diseases. Due to evolutionary conditions in living beings, among other functions have been developed and selected defense systems against the deleterious action of free radicals. Such systems are intrinsic in cells (at level intracellular and extracellular) and act together with the dietary exogenous antioxidants. All these antioxidant systems have very important role in preserving the oxide/reduction equilibrium in the cell. To understand the role of the transcription factor Nrf2 in regulating the processes of antioxidant defense, it must also know the role of many of the endogenous antioxidants that occur because of its activation. Therefore, this chapter makes a literature review of the most important general aspects of endogenous antioxidant systems, which will provide another point of view from which to approach the study and treatment of many chronic degenerative diseases, such as diabetes, hypertension, and Parkinson.

Keywords: oxidative stress, endogenous antioxidants, free radicals, catalase, glutathione

1. Introduction

The aerobic organisms use mitochondria as the main generator of energy for the realization of its vital functions. To do this, these organelles produce ATP through reactions of oxidation and reduction and attach to the tricarboxylic acid cycle with the electron transport chain. This happen due to to the oxidation of the food and of the NADH and FADH2, produced in different metabolic pathways, such as glycolysis, β-oxidation, and the same Krebs cycle. However, these reactions invariably result in the generation of reactive oxygen species (ROS) compounds that are unstable by having final layer of electrons unpaired and that, in trying to stabilize itself sequester electrons from other biomolecules, making them also destabilizes and, therefore, is no longer able to perform their duties properly, thus altering the homeostasis and, ultimately, causing cell death. Due to the current oxidant characteristics of the atmosphere on our planet, organisms are affected by imbalances in the oxidation-reduction reactions, not only on many of their metabolic reactions but also on external factors, such as microbial infections, xenobiotics, toxins from the diet, radiation, environmental pollution, and so on. All this in conjunction can contribute to the generation or aggravation of many diseases, such as cancer, diabetes, Parkinson and so on [1]. Other authors theorize that this imbalance in redox reactions has worked as an evolving pressure in order to develop effective mechanisms to eliminate the oxygen toxicity; this allowed the evolution of higher forms of living organisms, which are much more specialized and protected against negative actions of ROS (**Figure 1**) [2, 3].

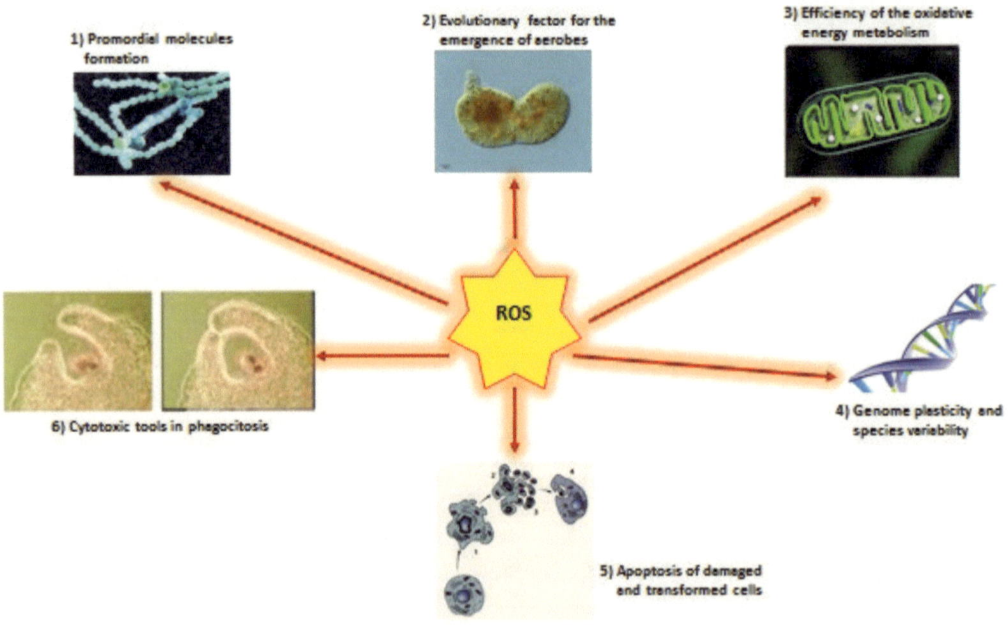

Figure 1. Evolving pressure of ROS in evolution of life in oxygen-rich atmosphere [2].

http://dx.doi.org/10.5772/65715

The reactive oxygen species can be either endogenous or exogenous [4, 5]. The transport chain of mitochondrial electron is the main source of endogenous ROS; the reduction in O_2 to H_2O_2 is carried out in four steps during which occur ROS and are as follows:

1. $O_2 + e \rightarrow O_2^{\circ}$ Superoxide radical

2. $O_2^{\circ} + H_2O \rightarrow H_2O^{\circ}$ Hydroperoxyl radical

3. $H_2O^{\circ} + e + H \rightarrow H_2O_2$ Hydrogen peroxide

4. $H_2O_2 + e \rightarrow OH^{\cdot} + OH^{\circ}$ Hydroxyl radical

Thus, these ROS being unstable, seek its stabilization capturing electrons from other biomolecules, altering its function, and therefore, strategies have been developed to maintain low concentrations of these ROS, thanks to the activity of endogenous antioxidant agents that may be of a protein or nonprotein nature. In **Figure 2**, it is described in a general way, the way in which they can generate ROS from the mitochondrial respiratory chain and it's debugging by some endogenous antioxidants [6–9].

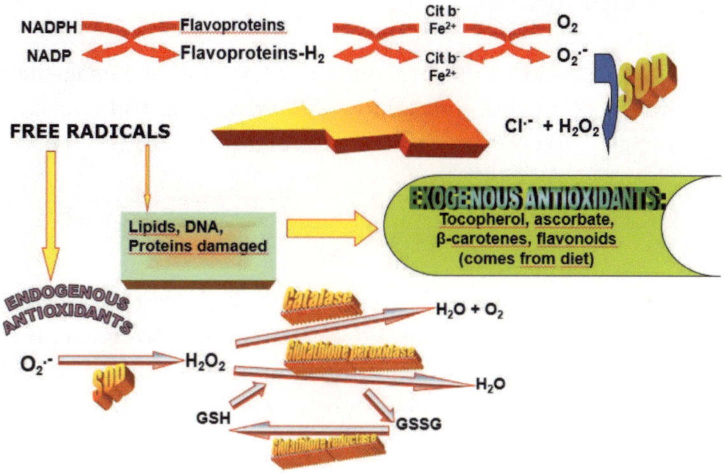

Figure 2. Formation of free radicals from the mitochondrial respiratory chain (modified from Ref. [10]).

Biomolecules in living organisms are highly exposed to oxidative stress, which is the main cause of damage to nucleic acids, proteins, carbohydrates, and polyunsaturated lipids, which finally develops cells mortality [11]. Reactive species derived from molecular oxygen (ROS) and nitrogen (RNS) have been deeply studied and new radical species such as chlorine (RCS), bromine (RBS), and sulfur-derived species have also been identified.

The "Free radicals" are molecular compounds with one-electron deficiency also denominated impaired electron in their outer orbital, examples of ROS are superoxide anion, hydroxyl radical, and hydrogen peroxide; nitric oxide and peroxynitrite are included in RNS.

Curiously, free radicals while having important chemical differences share similar mechanisms for damage at the level of biomolecules [11]. The oxidation of amino acid residues, the subsequent formation of protein aggregates by cross-linking and the production of protein fragments may result in the loss of activity and inactivation of enzymes and metabolic pathways, finally ending up with cell death [12].

Some authors have reported that at a physiological level there is a relation between the inactivation mechanisms by antioxidant system and the generation of ROS. This is related with a higher production of ROS when an organism presents harmful conditions, resulting in high oxidative stress conditions. In addition, if ROS are accumulated, the endogenous antioxidant defenses will not be enough. And immediately, oxidative modification in cellular membrane or intracellular molecules is performed in order to equilibrate the ROS antioxidant defense mechanisms [13].

In this chapter, a brief updated description is made of the main endogenous antioxidants, such as glutathione, lipoic acid, bilirubin, ferritin, superoxide dismutase, catalase, glutathione peroxidase, among others, as well as their participation in various pathological processes.

2. Endogenous nonprotein antioxidants

Four well-known main antioxidant mechanisms against oxidative damage have been largely studied (1) sequestration of transition metal ions, (2) scavenging and quenching of ROS and RNS, (3) ending of chain reactions by free radicals, and (4) molecular repairing of radical's damages.

2.1. Glutathione

GSH (L-γ-glutamyl-L-cysteinyl-glycine) is a non-protein thiol that reaches millimolar concentrations in most cell types. Its reduced form (GSH) is biologically active. It functions as an antioxidant defense against reactive oxygen/nitrogen species (ROS/RNS) as also with detoxication enzymes like GSH peroxidases and GSH-S-transferases [14, 15]. The GSH/glutathione disulfide is the major redox couple in animal cells [16].

Mitochondrial protection is exerted by GSH versus radicals and oxidant species by the contribution of a group of nutrients that can directly or indirectly protect mitochondria from oxidative damage and improve mitochondrial function [17]. The protection mechanism of these molecules prevents the generation of oxidants, scavenging free radicals, or inhibiting oxidant reactivity. Other mechanism includes increasing cofactors of mitochondrial enzymes that increase the kinetic constant of enzyme activity, which represents a protecting mechanism from further oxidation. The activation of enzymes such as hemeoxygenase 1 and NAD(P)H:quinone oxidoreductase 1, neutralize ROS and increase mitochondrial biogenesis [18].

2.2. Alpha-lipoic acid (LA)

Alpha-lipoic acid (LA) can deliver antioxidant activity in nonpolar and polar mediums and present antioxidant effect in its oxidized (LA) and reduced (DHLA [dihydrolipoic acid]) forms [19]. LA can actuate its antioxidant effect in any subcellular compartment of the body [20], and it is effective in recharging enzymes in the mitochondria [21]. Diabetes mellitus and neurodegenerative diseases can be controlled with LA due to the antioxidant properties of lipoate/dihydrolipoate system, influencing the tissue concentration of the reduced forms of other antioxidants. However, some evidences indicate that lipoic acid might also counteract NF-kB (Nuclear factor kappa-light-chain-enhancer of activated B cells) activation trigged by oxygen shock [22].

2.3. Coenzyme Q

Coenzyme Q (CoQ) is a benzoquinone derivate localized in the mitochondrial respiratory chain as well as in other internal membranes. CoQ is directly involved in energy transduction and aerobic adenosine triphosphate (ATP) production because it transports electrons in the respiratory chain and couples the respiratory chain to oxidative phosphorylation [23]. This compound is considered as an endogenously synthesized lipid soluble antioxidant, present in all membranes. The protective effect is extended to lipids, proteins, and DNA mainly because of its close localization to the oxidative cellular events [24]. In the inner mitochondrial membrane, CoQ has at least four different functions such as a redox carrier, antioxidant, activator of uncoupling proteins, and being a factor influencing the permeability transition pore (PTP). Also, it is proposed that lysosome contains a NADH-dependent CoQ reductase involved in translocation of protons into the lysosomal lumen [24].

2.4. Ferritin

Ferritin is an iron-binding protein. Which consists of its cytosolic form of two subunits, termed H and L. Twenty-four ferritin subunits assemble to form the apoferritin shell. Each apoferritin molecule is sequestrating iron atoms [25]. The main function of ferritin is to limit Fe (II) available to participate in the generation of oxygen-free radicals (ROS). It is not surprising that oxidant stress activates multiple pathways of ferritin regulation. In addition, it is proposed that ferritin provocates gene and protein alterations that coordinately limit oxidant toxicity. Some studies had demonstrated that exposure to heme group causes ferritin synthesis in endothelial cells and concordantly reduced their cytotoxic response to hydrogen peroxide [26, 27].

2.5. Uric acid

Uric acid is an intermediate product of the purine degradation pathway in the cell. Uric acid is degraded further by the enzyme uricase but there is evidence that in humans and great apes, the uricase gene was inactivated during hominoid evolution [28]. According to different demonstrations, uric acid and albumin are the two major antioxidants in human plasma, contributing 24% and 33%, respectively, of the total antioxidant activity [28]. It is well established that high blood levels of uric acid in humans protects cardiac, vascular, and neural cells from oxidative injury [29, 30]. The ability of urate to scavenge oxygen radicals and protect the erythrocyte membrane from lipid oxidation was first described by Kellogg and Fridovich [31], and was characterized further by Ames et al. [32].

The antioxidant effects of uric acid have been proposed particularly in conditions such as multiple sclerosis, Parkinson's disease, and acute stroke [13, 33–35]. While chronic elevations in uric acid are associated with increased stroke risk [36, 37], acute elevations in uric acid may provide some antioxidant protection. As a demonstration of this, cultured rat hippocampal neuronal cells were protected from oxidative stress with uric acid [38], and administration of uric acid 24 hours prior to middle artery occlusion also attenuated brain injury induced by acute ischemia in rats [38]. Uric acid is an antioxidant mainly in a hydrophilic environment, which is probably a major limitation of the antioxidant function of uric acid.

2.6. Bilirubin

Heme oxygenase-1 (HO-1) is the enzyme that generates carbon monoxide, iron, and biliverdin using the heme fraction as a substrate [39], and biliverdin reductase that reduces biliverdin to bilirubin being this molecule the ending product of the heme degradation. Bile pigments are potent in vitro scavengers of free radicals [39] reinforcing the concept that HO-1 is a crucial inducible antioxidant system engaged to assist against oxidative injury and other forms of cellular stress. Bilirubin has a role in the prevention of ischemic injury in isolated hearts [40], attenuation of oxidative damage in cultured cells [41], and modulation of airway smooth muscle contractility [42]. Also, bilirubin defends neurons against hydrogen peroxide-mediated damage [43], where a redox cycle between biliverdin and bilirubin appears to amplify this protective effect [44]. Recently, administration of biliverdin in vivo demonstrates to protect rat kidney, liver, and gut from ischemia-reperfusion injury [45–47]. In vitro experiments gave evidence that bile pigments scavenge NO and NO-related species [48]. Epidemiological studies sustain a beneficial action of bilirubin against the development of cardiovascular disease and cancer [49, 50].

The protection of bilirubin against classic coronary heart disease risk factor was demonstrated by Troughton et al. [51] and an increase of serum total bilirubin level is associated with the decrease of peripheral arterial disease [52]. Elevated serum bilirubin concentration protects from different coronary and microvascular dysfunctions and possibly against coronary atherosclerosis [53].

3. Endogenous protein antioxidants

This chapter provides an overview of the three protein antioxidants (with enzymatic activity), which are the first line of defense against oxidative stress on the body: superoxide dismutase (SOD), catalase (CAT), and glutathione peroxidase.

3.1. Superoxide dismutase (SOD)

Superoxide dismutases (SODs) are a group of key enzymes functioning as the first line of antioxidant defense by virtue of the ability to convert highly reactive superoxide radicals (dismutation) into hydrogen peroxide and molecular oxygen [54]. They have identified four isozymes of superoxide dismutase: (i) SOD1 is a metalloprotein binding copper and zinc ions that are localized in the cytosol of the cell [55, 56], (ii) SOD2, localized in the mitochondria and it is associated with the manganese or iron ions [56, 57], (iii) SOD3, localization is extracellular and is also associated with the copper and zinc, although it has a high molecular weight [56, 58] and it has high affinity for heparin and heparin sulfates [59], and (iv) SOD4 associated with nickel and found in various aerobic bacteria found in soil of class of Streptomyces [60, 61].

SOD1 (associated Cu/Zn) is present in a lot of Gram-negative bacteria pathogens and eukaryotes [62], although protists appear to lack SOD1 [63]. Plants contain SOD1 in the cytosol and in the chloroplast [64] and also been found in plant peroxisomes [65]. Animal cells possess dimeric SOD1 in the cytoplasm and in the nucleus, the intermembrane space of mitochondria [66] and peroxisomes [67]. However, the precise intracellular location of SOD1 is responsive to the metabolic state of cells and tissues [68]. SOD2 may be associated to Mn, Fe, or two ions (Mn/Fe).

SOD1 (associated with Cu/Zn) requires Cu and Zn for its biological activity; the loss of Cu results in its complete inactivation and is the cause of multiple diseases in human and animals [69]. It has a molecular weight of about 32 kDa. This group of enzymes works together with glutathione peroxidase and catalase to convert the superoxide radical into hydrogen peroxide. This enzyme has also been identified as a cause of familiar forms of amyotrophic lateral sclerosis (ALS) due to copper homeostasis is altered. Indeed, the total amount of copper ions in the mouse spinal cord, a region the most affected by ALS, is significantly elevated by expressing SOD1 [70, 71].

SOD2 has a molecular weight of about 96 kDa [59]. The SOD4 associated with Ni was discovered in Streptomyces [72] but has also been found in some genera of actinobacteria and cyanobacteria [73]. There is evidence that SOD2's levels to be regulated by MAPK activity in vitro [74, 75], also substances such as anthocyanins, polyphenols, alkaloids, and phytoalexins, are responsible for the induction of SOD2 mRNA expression; SOD2 can be induced by some inflammatory cytokines [76, 77], including tumor necrosis factor (TNF)-α, which may compensate for the inflammatory effect. The valine-to-alanine substitution in SOD2 Ala-16Val single nucleotide polymorphism (SNP) decreases the transport efficiency of the enzyme into the mitochondria and modifies the antioxidant defense against ROS. This process is an important pathophysiological mechanism in development and progression of diabetes and its complications [78, 79].

The SOD3 enzyme has many physiological effects; there are studies that reported reduced cardiovascular damage by recombinant administration of SOD3 [80, 81]. In the lung, mice with decreased levels of SOD3 had a significantly shortened life span and experienced death associated with lung edema under conditions of hyperoxia [82].

3.2. Catalase (CAT)

Catalase is a tetrameric porphyrin-containing enzyme that is located mainly in peroxisomes. It catalyzes the conversion of H_2O_2 to water and molecular oxygen in two steps [59, 83, 84]:

Catalase-Fe (III) + H_2O_2→compound I

Compound I + H_2O_2→catalase-Fe (III) + $2H_2O$ + O_2

The presence of bound NADPH in each subunit may help protect the enzyme from being inactivated by H_2O_2 [83]; the highest activity of this enzyme appears to be in the liver and erythrocytes [85]. Some reports indicate that factors such as stress and brain-derived neurotrophic factor (BDNF) are involved in the antioxidant capacity of many antioxidants endogenous; to this respect, Hacioglu et al. [86] studied a murine model where mice with BNDF deficiency under stress conditions showed an increment in CAT enzyme activity with respect to stressed wild-type mice, indicating that the ability to scavenge free radicals was diminished; and this suggest that normal wild-type mice have a better stress tolerance than BNDF heterozygous mice [86]. Catalase along with other antioxidant enzymes have been considered as biomarkers of oxidative stress in various organs; for example, in streptozotocin-induced diabetic rats, hepatic levels of these enzymes are dramatically reduced, although treatment with various plants can ameliorate this effect [87, 88]. For several decades, it was established that the levels of many antioxidant enzymes, including catalase, decline with age [89]. To this respect, Casado et al. [90] reported that in elderly patients with ischemic disease, COPD, and other diseases typical of old age, SOD levels are increased, but CAT levels are decreased; a phenomenon which they interpreted as a compensatory effect to balance the antioxidant system.

3.3. Glutathione peroxidase (GPx)

This enzyme can exist in two forms: selenium-dependent and selenium-independent, each with different subunits and different active sites [84, 91]. GPx catalyzes the reduction of H_2O_2 or organic peroxide (ROOH) to water or alcohol [59, 92]; this process occurs in the presence of GSH, which is converted into GSSG (oxidized glutathione) during this reaction. The reaction has special significance in the protection of the polyunsaturated fatty acids located within the cell membranes where the enzyme functions as a part of a multicomponent antioxidant defense system within the cell [93]. There are four isoforms in humans, cytosolic and mitochondrial (GPx1), cytosolic (GPx2), extracellular (GPx3), and the phospholipid peroxide (GPx4) [91, 94, 95]. The kidney and liver are the organs with the highest amount of GPx [85]. It is known that there is a competition between GPx and Cat for H_2O_2 as a substrate [59]. It has been found that in many other organs and tissues, such as the dorsal root ganglion (GDR), the GPx is the first enzyme that is activated under high levels of EROS, indicating the importance of this enzyme as a first line of defense against stress oxidative [96, 97]. Furthermore, have been found associations in the levels of this enzyme with skin diseases such as vitiligo; Asian vitiligo patients showed lower levels of GPx than the controls, but no difference was shown between populations of Caucasian vitiligo patients and Asian vitiligo patients [98].

Recent studies have involved GPx4 in carcinogenic processes, including the ferroptosis (nonapoptotic form of cell death that can be triggered by small molecules, which inhibit the biosynthesis of glutathione or GPx4); it was found that inactivation of GPx4 is crucial for the ferroptosis development and that overexpression of this enzyme blocks the action of the RSL-3, small GTPases called RAS, which is attacked by oncogenic RAS selective lethal (RSL) molecules; however, how RSL3 bind GPx4 to inhibit its activity is not known [99, 100].

4. Conclusions

As the reader will notice, endogenous antioxidants work as one big system that complements its main constituents to maintain redox balance in the body. When ROS levels rise and threaten the homeostatic processes of the human body, endogenous antioxidants are activated. The majority of them are expressed when some factors, such as factor Nrf2, are activated. However, endogenous antioxidants also work together with exogenous antioxidants from diet to decrease levels of ROS.

Reactive oxygen species could have or have not harmful effects, but they can also play a role in signaling different growth factors. This conclusion is supported by the hypothesis that decreased levels of ROS may lead to degenerative diseases, generating an interesting concept that ROS must be regulated but not eradicated.

In this chapter, some functional generalities of endogenous antioxidants were reviewed, and some aspects of their activity were discussed under conditions of high ROS levels; from this it follows that, although all of them work together, perhaps protein antioxidants, which have enzymatic activity, such as superoxide dismutase, catalase, and glutathione peroxidase, constitute the first line of defense against the oxidative stress.

Acknowledgements

The authors wants to thank the ENCB-IPN for the support received for this work through the SIP projects 20160554 and 20160443

Author details

Tomás Alejandro Fregoso Aguilar[1*], Brenda Carolina Hernández Navarro[1] and
Jorge Alberto Mendoza Pérez[2]

*Address all correspondence to: fisiobiologo@hotmail.com

1 Departamento de Fisiología, Escuela Nacional de Ciencias Biológicas, Instituto Politécnico Nacional, Mexico City, Mexico

2 Departamento de Ingeniería en Sistemas Ambientales, Escuela Nacional de Ciencias Biológicas, Instituto Politécnico Nacional, Mexico City, Mexico

References

[1] Königsberg F.M. (2007). Nrf2: The history of a new transcription factor to respond to oxidative stress. REB. 26: 18–25.

[2] Juránek I., Nikitovic D., Kouretas D., Hayes A.W., and Tsatsakis A.M., (2013). Biological importance of reactive oxygen species in relation of difficulties of treating pathologies involving oxidative stress by exogenous antioxidants. Food Chem Toxicol. 61: 240–247.

[3] Scmitt F., Renger G., Friedrich T., Kreslavski V.D., Zharmukhamedov S.K., Los D.A., Kuznetsov V.V., and Allakhverdiev S.I. (2014). Reactive oxygen species: Re-evaluation of generation, monitoring and role in stress-signaling in phototrophic organisms. Biochim Biophys Acta. 1837: 835–848.

[4] Borg C. (1993). Oxygen free radicals and tissue injury. In: Tarr M. and Samson editors. Oxygen free radicals in tissue damage. 2nd ed. Academic Press, New York, 25–55.

[5] Ballantine J.D. (1994). Pathology of oxygen toxicity. Academic Press, New York, 15–27.

[6] Loschen G., Azzi A., Richter C., and Flohe L. (1974). Superoxide radicals as precursors of hydrogen peroxide. FEBS Lett. 42: 68–72.

[7] Fridovich I., (1983). Superoxide radicals: an endogenous toxicant. Ann Rev Toxicol Pharmacol. 23: 239–257.

[8] Halliwell B. and Gutteridge J.M.C. (1992). Free radicals, antioxidants and human disease: Where are we now?. J Lab Clin Med. 119: 598–620.

[9] Ivanova E. and Ivanov B. (2000). Mechanisms of extracellular antioxidant defend. Exp Path Parasitol. 4: 49–59.

[10] Muñoz S.J.L. (2009). Defensas antioxidants endógenas, In. Morales-Gonzalez, Fernández S.A.M., Bautista A.M., Vargas M.N., Madrigal-Santillán E.O. (eds.) Los antioxidantes y las enfermedades crónico defenerativas. Universidad Autónoma del Estado de Hidalgo, México, 93–118.

[11] Halliwell B. and Whiteman M. (2004). Measuring reactive species and oxidative damage in vivo and in cell culture: how should you do it and what do the results mean? Br J Pharmacol.; 142:231–255

[12] Stadman E.R. and Levine R.L. (2003). Free radical-mediated oxidation of free amino acids and amino acid residues in proteins. Amino Acids. 25:207–218.

[13] Ott M., Gogvadze V., Orrenius S. and Zhivotovsky B. (2007). Mitochondria, oxidative stress and cell death. Apoptosis. 2(5): 913–922.

[14] Lillig C.H., Berndt C., Vergnolle O., Lönn M.E., Hudemann C., Bill E. and Arne H. (2005). Characterization of human glutaredoxin 2 as iron–sulfur protein: a possible role as redox Sensor. Proc Natl Acad Sci USA. 102: 8168–8173.

[15] Jeon Ch, and Joo S.H. (2016). Downregulation of reactive oxygen species in apoptosis. J Cancer Prevention. 21: 13–20.

[16] Ardite E., Sans M., Panes J., Romero F.J., Pique J.M. and Fernandez-Checa J.C. (2000). Replenishment of glutathione levels improves mucosal function in experimental acute colitis. Lab Invest. 80(5):735–744.

[17] Liu J. and Ames B.N. (2005). Reducing mitochondrial decay with mitochondrial nutrients to delay and treat cognitive dysfunction, Alzheimer's disease and Parkinson's disease. Nutr Neurosci. 8(2):67–89.

[18] Liu J. (2008). The effects and mechanisms of mitochondrial nutrient alpha-lipoic acid on improving age-associated mitochondrial and cognitive dysfunction: an overview. Neurochem Res. 33(1):194–203.

[19] Goraca A., Huk-Kolega H., Piechota A., Kleniewska P., Ciejka E. and Skibska B. (2011) Lipoic acid—biological activity and therapeutic potential. Pharmacol Rep. 63:849–858. PMID: 22001972

[20] Packer L., Tritschler H. and Wessel K. (1997). Neuroprotection by the metabolic antioxidant alpha-lipoic acid. Free Radical Biol Med. 22:359–378. PMID: 8958163

[21] Arivazhagan P., Ramanathan K. and Panneerselvam C. (2001). Effect of DL-alpha-lipoic acid on mitochondrial enzymes in aged rats. Chem Biol Interact. 138:189–198. PMID: 11672700.

[22] Lee C.K., Lee E.Y., Kim Y.G., Mun S.H., Moon S.B. and Yoo B. (2008). Alpha-lipoic acid inhibits TNF-alpha induced NF-kappa B activation through blocking of MEKK1-MKK4-IKK signaling cascades. Int Immunopharmacol. 8(2):362–370.

[23] Rohr-Udilova N.V., Stolze K., Sagmeister S., Nohi H., Shulte-Herman R. and Grasi-Kraupp B. (2008). Lipid hydroperoxides from processed dietary oils enhance growth of hepatocarcinoma cells. Mol Nutr Food Res. 2008. 52(3):352–359.

[24] Gille L. and Nohl H. (2000). The existence of a lysosomal redox chain and the role of ubiquinone. Arch Biochem Biophys. 375:347–354.

[25] Nunomura A., Moreira P.I., Takeda A., Smith M.A. and Perry G. (2007). Oxidative RNA damage and neurodegeneration. Curr Med Chem. 14(28): 2968–2975.

[26] Pauwels E.K., Erba P.A. and Kostkiewicz M. (2007). Antioxidants: A tale of two stories. Drug News Perspect. 20(9):579–585.

[27] Balla G., Jacob H.S., Balla J., Rosenberg M., Nath K., Apple F., Eaton J.W. and Vercellotti G.M. (1992). Ferritin: a cytoprotective antioxidant strategem of endothelium. J Biol Chem. 267: 18148–18153.

[28] Miller N.J. and Evans C.A. (1996). Spectrophotometric determination of antioxidant activity. Redox Rep. 2:161–171.

[29] Shi Q. and Gibson G.E. (2007). Oxidative stress and transcriptional regulation in Alzheimer disease. Alzheimer Dis Assoc Disord. 21(4): 276–291.

[30] Stocker R. and Keaney J.F. (2004). Role of oxidative modifications in atherosclerosis. Physiol. Rev. 84: 1381–1478.

[31] Kellogg EW, Fridovich I. (1977). Liposome oxidation and erythrocyte lysis by enzymically generated superoxide and hydrogen peroxide. J Biol. Chem. 252: 6721–6728

[32] Ames BN, Cathcart R, Schwiers E, Hochstein P. (1981). Uric acid provides an antioxidant defense in humans against oxidant- and radical-caused aging and cancer: a hypothesis. Proc Natl. Acad. Sci. USA.78:.6858–6862.

[33] Nishikori M. (2005). Classical and alternative NF-kB activation pathways and their roles in lymphoid malignancies. J Clin Hematopathol. 45:15–24.

[34] Hooper D.C., Spitsin S., Kean R.B., Champion J.M., Dickson G.M., Chaudhry I. and Koprowski H. (1998). Uric acid, a natural scavenger of peroxynitrite, in experimental allergic encephalomyelitis and multiple sclerosis. Proc Natl Acad Sci. 95:675–680.

[35] Spitsin S.V., Scott G.S., Mikheeva T., Zborek A., Kean R.B., Brimer C.M., Koprowski H. and Hooper D.C. (2002). Comparison of uric acid and ascorbic acid in protection against EAE. Free Radic Biol Med. 33:1363–1371.

[36] André M. and Felley-Bosco E. (2003) Heme oxygenase-1 induction by endogenous nitric oxide: influence of intracellular glutathione. FEBS Lett. 546:223–227.

[37] Weir C.J., Muir S.W., Walters M.R. and Lees K.R. (2003). Serum urate as an independent predictor of poor outcome and future vascular events after acute stroke. Stroke. 34:1951–1956.

[38] Yu Z.F., Bruce-Keller A.J., Goodman Y. and Mattson M.P. (1998). Uric acid protects neurons against excitotoxic and metabolic insults in cell culture, and against focal ischemic brain injury in vivo. J Neurosci Res. 53:613–625.

[39] Stocker R., Yamamoto Y., McDonagh A., Glazer A.N. and Ames B.N. (1987). Bilirubin is antioxidant of possible physiological importance. Science. 235:1043–1046.

[40] Clark J.E., Foresti R., Sarathchandra P. Kaur H., Green C.J. and Motterlini R. (2000). Heme oxygenase-1-derived bilirubin ameliorates postischemic myocardial dysfunction. Am J Physiol Heart Circ Physiol. 278: H643–H651.

[41] Foresti R., Sarathchandra P., Clark J.E., Green C.J. and Motterlini R. (1999). Peroxynitrite induces haem oxygenase-1 in vascular endothelial cells: a link to apoptosis. Biochem J. 339:729–736.

[42] Samb A., Taille C., Almolki A. Mégret J., Staddon J.M, Aubier M. and Boczkowski J. (2002). Heme oxygenase modulates oxidant-signaled airway smooth muscle contractility: role of bilirubin. Am J Physiol Lung Cell Mol Physiol. 283: L596–L603.

[43] Dore S. and Snyder S.H. (1999). Neuroprotective action of bilirubin against oxidative stress in primary hippocampal cultures. Ann NY Acad Sci. 890:167–172.

[44] Baranano D.E., Rao M., Ferris C.D. and Snyder S.H. (2002). Biliverdin reductase: a major physiologic cytoprotectant. Proc Natl Acad Sci. 99:16093–16098.

[45] Kato Y., Shimazu M., Kondo M., Uchida K., Kumamoto Y., Wakabayashi G., Kitajima M. and Suematsu M. (2003). Bilirubin rinse: a simple protectant against the rat liver graft injury mimicking heme oxygenase-1 preconditioning. Hepatology 38:364–373.

[46] Adin C.A., Croker B.P. and Agarwal A. (2005). Protective effects of exogenous bilirubin on ischemia-reperfusion injury in the isolated, perfused rat kidney. Am J Physiol Renal Physiol. 288: F778–F784.

[47] Nakao A., Otterbein L.E., Overhaus M., Sarady J.K., Tsung A., Kimizuka K., Nalesnik M.A., Kaizu T., Uchiyama T., Liu F., Murase N., Bauer A.J. and Bach F.H. (2004). Biliverdin protects the functional integrity of a transplanted syngeneic small bowel. Gastroenterology 127:595–606.

[48] Kaur H., Hughes M.N., Green C.J., Naughton P., Foresti R. and Motterlini R. (2003). Interaction of bilirubin and biliverdin with reactive nitrogen species. FEBS Lett. 543:113–119.

[49] Hopkins P.N., Wu L.L., Hunt S.C., James B.C., Vincent G.M., and Williams R.R. (1996). Higher serum bilirubin is associated with decreased risk for early familial coronary artery disease. Arterioscler Thromb Vasc Biol. 16:250–255.

[50] Temme E.H., Zhang J., Schouten E.G., Zhang J.J. and Kesteloot H. (2001). Serum bilirubin and 10-year mortality risk in a Belgian population. Cancer Causes Control. 12:887–894.

[51] Troughton J., Woodside J.V. and Young I.A. (2007). Bilirubin and coronary heart disease risk in the Prospective Epidemiological Study of Myocardial Infarction (PRIME). Eur J Cardiovasc Prev Rehabil. 14(1):79–84.

[52] Perlstein T.S., Pande R.L., Beckman J.A. and Creager M.A. (2008). Serum total bilirubin level and prevalent lower-extremity peripheral arterial disease: National Health and Nutrition Examination Survey (NHANES) 1999 to 2004. Arterioscler Thromb Vasc Biol. 28(1):166–172.

[53] Gullu H., Erdogan D., Tok D., Topcu S., Caliskan M., Ulus T. and Muderrisoglu H. (2005). High serum bilirubin concentrations preserve coronary flow reserve and coronary microvascular functions. Arterioscler Thromb Vasc Biol. 25(11):2289–2294.

[54] Wang W., Xia M.X., Chen J., Yuan R., Deng F.N. and Shen F.F. (2016). Gene expression characteristics and regulation mechanisms of superoxide dismutase and its physiological roles in plants under stress. Biochemistry (Mosc.) 81 (5): 465–480.

[55] McCord J.M. and Fridovich I. (1969). Superoxide dismutase: an enzymic function for erythrocuprein (hemocuprein). J Biol Chem. 244 (22): 6049–6055.

[56] Laukkanen M.O. (2016). Extracellular superoxide dismutase: growth promoter or tumor suppressor?. Oxid Med Cell Longev. 16: 1–9. doi:10.1155/2016/3612589

[57] Weisiger R.A. and Fridovich I. (1973). Superoxide dismutase. Organelle specificity. J Biol Chem. 248 (10): 3582–3592.

[58] Marklund S.L. (1982). Human copper-containing superoxide dismutase of high molecular weight. Proc Natl Acad Sci 79 (24 I): 7634–7638.

[59] Valko M., Rhodes C.J., Moncol, J., Izacovic M. and Mazur M. (2006). Free radicals, metals and antioxidants in oxidative stress-induced cancer. Chem Biol Interact. 160: 1–40.

[60] Wuerges J, Lee J.W., Yim Y.I., Yim H.S., Kang S.O. and Carugo K.D. (2004). Crystal structure of nickel-containing superoxide dismutase reveals another type of active site. Proc Natl Acad Sci USA 101: 8569–8574.

[61] Anju A., Jeswin J., Thomas P.C., Paulton M.P. and Vijayan K.K. (2013). Molecular cloning, characterization and expression analysis of cytoplasmic Cu/Zn-superoxide dismutase (SOD) from pearl oyster Pinctada fucata. Fish Shellfish Immunol. 34: 943–950.

[62] Ammendola S., Pasquali P., Pacello F., Rotilio G., Castor M, Libby, S.J., Figueroa-Bossi N., Bossi L., Fang F.C., and Battistoni A. (2008). Regulatory and structural differences in the Cu,Zn-superoxide dismutases of salmonella enteric and their significance for virulence. J Biol Chem. 283 (20): 13688–13699.

[63] Wilkinson S.R., Prathalingam S.R., Taylor M.C., Ahmed A., Horn D. and Kelly J.M. (2006) Functional characterization of the iron superoxide dismutase gene repertoire in *Trypanosoma brucei*. Free Radic. Biol. Med. 40, 193–195.

[64] Fink R.C. and Scandalios J.G. (2002). Molecular evolution and structure-function relationships of the superoxide dismutase gene families in angiosperms and their relationship to other eukaryotic and prokaryotic superoxide dismutases. Arch Biochem Biophys. 399: 19–36.

[65] Bueno P., Varela J., Giménez-Gallego G. and Del Río L.A. (1995). Peroxisomal copper, zing superoxide dismutase - characterization of the isoenzyme from watermelon cotyledons. Plant Physiol. 108: 1151–1160.

[66] Chang L.Y., Slot J.W., Geuze H.J. and Crapo J.D. (1988) Molecular immunocytochemistry of the CuZn superoxide dismutase in rat hepatocytes. J Cell Biol. 107: 2169–2179.

[67] Kira Y., Sato E.F. and Inoue M. (2002). Association of Cu, Zn-type superoxide dismutase with mitochondria and peroxisomes. Arch Biochem Biophys. 399: 96–102.

[68] Geller B.L. and Winge D.R. (1982). Rat liver Cu, Zn-superoxide dismutase. J Biol Chem. 257, 8945–8952.

[69] Noor R., Mittal S. and Iqbal J. (2002). Superoxide dismutase-applications and relevance to human diseases. Med Sci Monit. 8: 210–215.

[70] Tokuda E., Okawa E., Watanabe S., Ono S. and Marklund S.L. (2013). Dysregulation of intracellular copper homeostasis is common to transgenic mice expressing human mutant superoxide dismutase-1s regardless of their copper-binding abilities. Neurobiol Dis. 54: 308–319.

[71] Tokuda E. and Furukawa Y. (2016). Copper homeostasis as a therapeutic target in amyotrophic lateral sclerosis with SOD1 mutations. Int J Mol Sci. 17 (636): doi: 10.3390/ijms17050636.

[72] Youn H.-D., Kim E.J., Roe J.H., Hah Y.C. and Kang, S.O. (1996) A novel nickel-containing superoxide dismutase from Streptomyces spp. Biochem J. 318: 889–896.

[73] Schmidt A., Gube M., Schmidt A., and Kothe E. (2009). In silico analysis of nickel containing superoxide dismutase evolution and regulation. J Basic Microbiol. 49: 109–118.

[74] Hashimoto N., Noda T., Kim S.J., Yamauchi H., Takigawa S., Matsuura-Endo C., Suzuki T., Han K.H. and Fukushima M. (2010). Colored potato extracts induce superoxide dismutase-2 mRNA via ERK1/2 pathway in HepG2 cells. Plant Foods Hum Nutr. 65: 266–270.

[75] Han K.H., Hashimoto N. and Fukushima M. (2016). Relationships among alcoholic liver disease, antioxidants, and antioxidants enzymes. World J Gastroenterol. 22 (1): 37–49.

[76] Guo Z., Boekhoudt G.H. and Boss J.M. (2003) Role of the intronic enhancer in tumor necrosis factor-mediated induction of manganous superoxide dismutase. J Biol Chem. 278: 23570–23578.

[77] Borras C., Gambini J., Gomez-Cabrera M.C., Sastre J., Pallardo F.V., Mann G.E. and Vina J. (2005). 17beta-oestradiol up-regulates longevity related, antioxidant enzyme expression via the ERK1 and ERK2 [MAPK]/NFkappaB cascade. Aging Cell 4: 113–118.

[78] Ascencio-Montiel I.J., Parra E.J., Valladares-Salgado A., Gomez-Zamudio J.H., Kumate-Rodriguez J., Escobedo-dela-Pena J., and Cruz M. (2013). SOD2 gene Val16Ala polymorphism is associated with macroalbuminuria in Mexican type 2 diabetes patients: a comparative study and meta-analysis. BMC Med Genet. 14:110. doi: 10.1186/1471-2350-14-110.

[79] Pourvali K., Abbasi M., and Mottaghi A. (2016). Role of superoxide dismutase 2 gene Ala16Val polymorphism and total antioxidant capacity in diabetes and its complications. Avicenna J Med Biotech. 8(2): 48–56.

[80] Sjoquist P.O. and Marklund S.L. (1992). Endothelium bound extracellular superoxide dismutase type C reduces damage in reperfused ischaemic rat hearts. Cardiovasc Res. 26(4): 347–350.

[81] Leite P.F., Danilovic A., Moriel P., Dantas K., Marklund S., Dantas A.P.V., and Laurindo F.R.M. (2003). Sustained decrease in superoxide dismutase activity underlies constrictive remodeling after balloon injury in rabbits. Arterioscler Thromb Vasc Biol. 23: 2197-2202.

[82] Carlsson L.M., Jonsson J., Edlund T. and Marklund S.L. (1995). Mice lacking extracellular superoxide dismutase are more sensitive to hyperoxia. Proc Natl Acad Sci. 92(14): 6264–6268.

[83] Chaudiere J. and Frerrari-LLiou R. (1999). Intracellular antioxidants: from chemical to biochemical mechanisms. Food Chem Toxicol. 37: 949–962.

[84] Aslani B.A. and Ghobadi S. (2016). Studies on oxidants and antioxidants with a brief glance at their relevance to the immune system. Life Sci. 146: 163–173.

[85] Young I.S. and Woodside J.V. (2001). Antioxidants in health and disease. J Clin Pathol. 54: 176–186.

[86] Hacioglu G., Senturk A., Ince I. and Alver A. (2016). Assessment of oxidative stress parameters of brain-derived neurotrophic factor heterozygous mice in acute stress model. Iran J Basic Med Sci. 19 (4): 388–393.

[87] Ayeleso A., Brooks N. and Oguntibeju O. (2014). Modulation of antioxidant status in streptozotocin-induced diabetic male Wistar rats following intake of red palm oil and/or rooibos. Asian Pac J Trop Med. 7(7): 536–544.

[88] Oguntibeju O., Meyer S., Aboua Y.G. and Goboza M. (2016). Hypoxis hemerocallidea significantly reduced hyperglycaemia and hyperglycaemic-induced oxidative stress in the liver and kidney tissues of streptozotocin-induced diabetic male Wistar rats. Evid Based Complement Alternat Med. 2016: 1–10

[89] Guemouri L., Arthur Y., Herberi B., Jeandel C., Cuny G. and Siest G. (1991). Biological variability of superoxide dismutase, glutathione peroxidase and catalase in blood. Clin Chem. 37: 1932–1937.

[90] Casado A., De la Torre R., López-Fernández E., Carrascosa D. and Venarucci D. (1998). Levels of superoxide dismutase and catalase in diseases of the elderly. Gac Méd Méx. 134 (5): 539–544.

[91] Rahman K. (2007). Studies on free radicals, antioxidants, and co-factors. Clin Interv Aging. 2: 219–236.

[92] Dickinson D.A. and Forman H.J. (2002). Cellular glutathione and thiols metabolism. Biochem Pharmacol. 64: 1019–1026.

[93] Gathwala G. and Aggarwal R. (2016). Selenium supplementation for the preterm indian neonate. Indian J Public Health. 60: 142–144.

[94] Forsberg L., De Faire U. and Morgenstern R. (2001). Oxidative stress, human genetic variation, and disease. Arch Biochem Biophys. 389 84–93.

[95] Jan A.T., Azam M., Siddiqui K., Ali A., Choi I. and Haq Q.M.R. (2015). Heavy metals and human health: mechanistic insight into toxicity and counter defense system of antioxidants. Int J Mol Sci. 5(16): 29592–29630.

[96] Halliwell B. and Gutteridge J.M.C. (2015). Free radicals in biology and medicine. 5th ed. Oxford University Press. 753 pp.

[97] Duggett N.A., Griffiths L.A., McKenna O.E., De Santis V., Yongsanguanchai N., Mokori E.B. and Flatters S.J.L. (2016). Oxidative stress in the development, maintenance and resolution of paclitaxel-induced painful neuropathy. Neuroscience.333: 13–26

[98] Xiao B.H., Shi M., Chen H., Cui S., YanWu Y., Gao X.H., and Chen H.D. (2016). Glutathione peroxidase level in patients with vitiligo: a meta-analysis. BioMed. Res. Int. 2016: 1–11

[99] Yang W.S. and Stockwell B.R. (2008). Synthetic lethal screening identifies compounds activating iron-dependent, nonapoptotic cell death in oncogenic-RAS-harboring cancer cell. Chem Biol. 15: 234–245.

[100] Cao J.Y. and Dixon S.J. (2016). Mechanisms of ferroptosis. Cell Mol Life Sci. 73: 2195–2209.

Chapter 2

NRF2 Rewires Cellular Metabolism to Support the Antioxidant Response

Ting-Yu Lin, Lewis C. Cantley and Gina M. DeNicola

Additional information is available at the end of the chapter

http://dx.doi.org/10.5772/65141

Abstract

The transcription factor (nuclear factor-erythroid 2 p45-related factor 2, NRF2) is a master regulator of the cellular response to oxidative insults. While antioxidant response enzymes are well-characterized transcriptional targets of NRF2, it is recently becoming clear that NRF2 also supports cellular detoxification through metabolic rewiring to support the antioxidant systems. In this chapter, we discuss the regulation of NRF2 and how NRF2 activation promotes the antioxidant defense of cells. Furthermore, we discuss how reactive oxygen species influence cellular metabolism and how this affects antioxidant function. We also discuss how NRF2 reprograms cellular metabolism to support the antioxidant response and how this functions to funnel metabolic intermediates into antioxidant pathways. This chapter concludes by exploring how these factors may contribute to both normal physiology and disease.

Keywords: NRF2, KEAP1, antioxidant response, metabolism, glucose, glutamine, ROS

1. Introduction

The transcription factor (nuclear factor-erythroid 2 p45-related factor 2, NRF2) is a master regulator of the cellular response to oxidative insults. While antioxidant response enzymes are well-characterized transcriptional targets of NRF2, it is recently becoming clear that NRF2 also supports cellular detoxification through metabolic rewiring to support the antioxidant systems. Reactive oxygen species (ROS) are highly reactive oxygen-containing molecules, including hydrogen peroxide (H_2O_2), superoxide, and hydroxyl-free radical. The production of ROS comes from a variety of organelles within the cell, including the mitochondria, the endoplasmic reticulum (ER), and peroxisomes. Mitochondria are believed to be the major source of intracellular ROS, as this is where active respiration takes place, which steadily converts oxygen to superoxide [1]. The control of ROS levels in cells is crucial for cellular homeostasis, as Moderate ROS levels in cells can serve as signaling molecules, but excessive ROS production can lead to damage to DNA, proteins, and lipids [2]. Therefore, redox homeostasis is important for cell survival and is achieved by the balance of ROS production and elimination. Several factors have been suggested to affect ROS levels in cells, including hypoxia, ER stress, metabolic alteration, and oncogenes [3].

It has been shown that hypoxia can stimulate ROS production in mitochondria, which activates hypoxia-inducible transcription factor 1 alpha (HIF1α) [4]. Similarly, misfolded proteins that cause ER stress can elicit the unfolded protein response (UPR) and promote ROS accumulation [5]. In addition, metabolic alterations were thought to be a source of oxidative stress, as metabolic reactions are often accompanied by ROS production. Metabolic alterations have often been found in cancer cells, as cancer cells have a higher demand for ATP and other metabolites to sustain their unlimited proliferation and growth, and the higher metabolic activity in cells would lead to more ROS generation due to more mitochondria respiration. Furthermore, oncogenes are originally believed to elevate the ROS levels in cells, supported by the facts that ectopic expression of different oncogenes (e.g., Myc and KRAS) increase ROS levels in cells [6, 7]. However, a more recent study observed that physiological levels of oncogene (KRAS^{G12D} and MYC) expression could trigger a NRF2-dependent antioxidant response to downregulate ROS levels in cells [8]. These different results could be accounted for by both the expression of NADPH oxidase induced by KRAS^{G12D} overexpression, which leads to ROS generation, and an increase in NRF2-dependent ROS scavenging in cells expressing physiological levels of KRAS^{G12D}. In this chapter, we will be discussing the antioxidant programs regulated by NRF2 and how NRF2 supports ROS detoxification through metabolic rewiring. Finally, we will be discussing how these pathways contribute to the physiological function and diseases.

1.1. NRF2 and KEAP1

Nuclear factor-erythroid 2 p45-related factor 2 (NRF2 or NFE2L2) is a stress-responsive cap'n'collar (CNC) basic region leucine zipper (bZIP) transcription factor that directs various transcriptional programs in response to oxidative stress. Kelch-like ECH-associated protein 1 (Keap1) is believed to be the major repressor of NRF2, supported by the evidence that disruption of *Keap1* in the mouse increased the abundance and activity of NRF2 [9]. Under basal conditions, NRF2 is kept inactive through binding to its negative regulator KEAP1, which is a redox-regulated substrate adaptor for the Cullin (Cul)3-RING-box protein (Rbx)1 ubiquitin ligase complex that directs NRF2 for degradation [10]. Keap1 is a cysteine-rich protein that can be oxidized by ROS and results in conformational change, which in turn liberates NRF2. The free NRF2 translocates into the nucleus and heterodimerizes with Maf proteins to bind to a specific DNA sequence called an antioxidant response element (ARE, 5′-TGACNNNGC-3′) to activate the expression of a group of detoxifying and antioxidant genes, such as *glutathione S-transferase* (*GST*) and *NAD(P)H:quinone oxidoreductase-1* (*NQO1*) [11].

http://dx.doi.org/10.5772/65141

It is generally accepted that NRF2 is sequestered by KEAP1 in the cytoplasm under normal conditions and is released from KEAP1 before translocating into the nucleus under oxidative stress (**Figure 1**). Additionally, it has been demonstrated that both modification on KEAP1 at cysteine 151 and PKC-delta-mediated NRF2 phosphorylation are required for the release of NRF2 from KEAP1 [12]. In addition, there are several other stress kinases that were found to regulate NRF2 nuclear localization by phosphorylation, including the ER stress kinase PERK and the energy sensor AMPK [13, 14]. However, there is also evidence that KEAP1 is not only restricted to the cytoplasm but can undergo nuclear-cytoplasmic shuttling. A study found that the NRF2-KEAP1 complex keeps shuttling constantly between the cytoplasm and the nucleus, and in the nucleus, the nuclear protein prothymosin α competes for NRF2 binding to KEAP1 and enables the liberated NRF2 to activate NRF2 target genes [15]. Additionally, the nuclear export signal (NES) in KEAP1 has been shown to be important for NRF2 regulation, as it was found to facilitate the transport of the NRF2-KEAP1 complex out of the nucleus, which terminates NRF2-mediated antioxidant signaling after stress [16]. Thus, KEAP1 may regulate NRF2 localization through multiple mechanisms, and the model is being constantly refined.

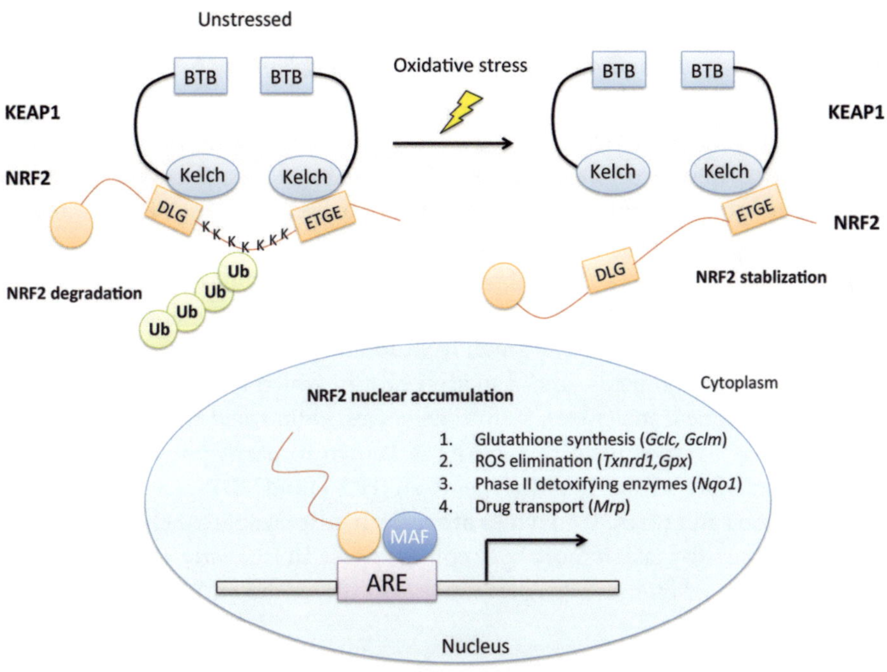

Figure 1. KEAP1-mediated NRF2 regulation. Under normal condition, NRF2 binds to KEAP1 dimers through the DLG and ETGE motifs within the Neh2 domain in NRF2, which leads to the constitutive ubiquitination and proteasomal degradation of NRF2. ROS, which induces oxidative stress, can oxidize the Cys residues on KEAP1, leading to the conformational change of KEAP1 and thus releasing NRF2. NRF2 can translocate into the nucleus, heterodimerize with MAF proteins, and bind to the specific DNA sequence termed antioxidant response element (ARE) to activate the downstream antioxidant genes, including the enzymes required for GSH production, ROS elimination, phase II metabolism, and the export of xenobiotics.

Recently, there is a theory explaining how NRF2 is stabilized under stress conditions, called the "two-site substrate hinge and latch model." It is known that each of the two dimeric KEAP1 molecules will bind NRF2 separately through their respective Kelch repeat domain to either the ETGE or DLG motifs in NRF2. It is proposed that the dimeric KEAP1 captures NRF2 first through the ETGE motif, which has 100-fold more affinity to KEAP1, followed by DLG motif docking onto the other Kelch repeat domain. The stress-induced KEAP1 conformational change leads to the release of the DLG motif from KEAP1, preventing NRF2 from ubiquitination by Cul3. It is proposed that the non-ubiquitinated NRF2 would remain bound to the KEAP1-Cul3 complex and therefore prevent the newly translated NRF2 from inhibition by the E3-ubiquitin complex, thereby promoting NRF2 accumulation in the nucleus [17]. In addition, the other mechanisms have been reported to stabilize NRF2 through interfering with the formation of the NRF2-KEAP1 complex. Those KEAP1-binding competitors, such as $p21^{Cip1/WAF1}$ and BRCA1, were found interacting with NRF2 and thus block the association of KEAP1 with NRF2 [18, 19]. Interestingly, another mechanism was proposed as an alternative way to stabilize NRF2 through the interaction between the autophagy adaptor protein p62/sequestosome-1 (p62/SQSTM1) and KEAP1, which leads to the sequestration of KEAP1 in the autophagy-deficient cells and enables NRF2 accumulation [20].

2. The antioxidant response program regulated by NRF2

The elimination of xenobiotics from cells is an important process for preventing cell toxicity and can be divided into three stages: phases I, II, and III. Phase I involves a transfer of a hydroxyl, a carboxyl, or an amino group to the toxic compound, which is often mediated by the cytochrome P450 enzymes. The modified metabolites generated by phase I enzymes will further be modified by phase II conjugation enzymes, which are mostly transferase enzymes that can transfer the endogenous hydrophilic molecules to the metabolite to increase the solubility and promote excretion. The small hydrophilic molecules include glucuronic acid, glutathione, sulfate, amino acids, and a methyl group, which are catalyzed by glucuronyl transferases, glutathione transferases, sulfotransferases, amino acid transferases, and N- and O-methyltransferases, respectively [21]. NRF2 is known to participate in the regulation of several phase II enzymes, such as heme oxygenase-1 (HO-1) and UDP-glucuronosyltransferase (UGT). While phase I and phase II enzymes are majorly responsible for chemically modifying the hydrophobic toxin to make it more hydrophilic, phase III is mostly related to the function of drug efflux performed by those drug transporters, such as the multidrug resistance (MDR) pump.

NRF2 coordinates with several antioxidant pathways in response to oxidative stress (**Figure 2**). The best-known example is the regulation of the synthesis and regeneration of glutathione (GSH), which is the most abundant antioxidant molecule in cells. GSH production is supported by NRF2 through controlling several enzymes responsible for GSH synthesis, such as the glutamate-cysteine ligase (GCL) complex composed of two subunits, the modifier (GCLM) subunit and the catalytic (GCLC) subunit, which catalyze the rate-limiting step of GSH synthesis that converts cysteine and glutamate into GSH [22]. NRF2 also controls the expression of glutathione S-transferases (GST) and glutathione peroxidases 2 (GPX2), which are enzymes known for detoxifying epoxide and H_2O_2, respectively. Additionally, NRF2 also regulates the expression of the solute carrier family 7 member 11 (SLC7A11), which encodes the cystine/glutamate transporter XCT [23]. XCT can import cystine (CySS) into the cell, which can be reduced to cysteine (Cys) using GSH or thioredoxin reductase 1 (TXNRD1). Cys can then be used to support GSH production. GSH is utilized to convert H_2O_2 to H_2O, accompanied by the production of oxidized glutathione (GSSG), which can be reduced back to GSH through glutathione reductase 1 (GSR1). This step is particularly important as cells need to replenish the pool of reduced GSH to keep carrying on the next reduction cycle [24].

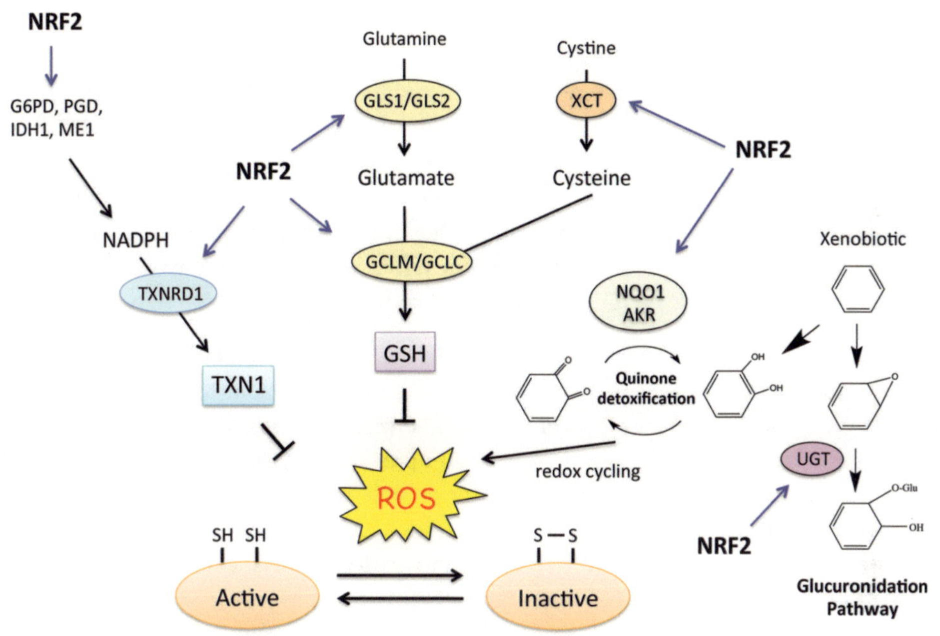

Figure 2. The antioxidant response program regulated by NRF2. NRF2 is responsible for initiating several different antioxidant pathways to eliminate ROS, including promoting the synthesis of glutathione (GSH) and thioredoxin (TXN). GSH, the most abundant antioxidant molecule in cells, derived from cysteine and glutamate, can eliminate ROS through the reaction mediated by glutathione peroxidase (GPX) and glutathione S-transferase (GST), and the oxidized GSH is regenerated by NADPH through glutathione reductase (GSR). Similarly, TXN is regenerated by thioredoxin reductase (TXNRD), which also requires NADPH. Glutamate and cystine are the two components required for GSH synthesis, which can be converted to GSH through the action of the glutamate-cysteine ligase modifier (GCLM) subunit and the GCL catalytic (GCLC) subunit. The cystine/glutamate transporter XCT transports cystine into the cell which is reduced to cysteine via TXNRD, and glutamate is generated by glutaminase 1 (GLS1) and GLS2 using glutamine as a substrate. Furthermore, xenobiotics undergo oxidation, reduction, or hydrolysis by cytochrome P450 enzymes and are subsequently conjugated with glucuronic acid, catalyzed by UDP-glucuronosyltransferase (UGT), and exported out of the cells by multidrug resistance-associated proteins (MRP). Some of the xenobiotics such as quinones can undergo redox cycling which results in ROS production, which is counteracted by the NADPH-required reduction reactions catalyzed by aldo-keto reductase (AKR) and NAD(P)H: quinone oxidoreductase 1 (NQO1). The role of NRF2 in mediating the antioxidant response is through (1) regulation of NADPH-generating enzymes (G6PD, PGD, IDH1, and ME1); (2) upregulation of the enzymes required for GSH production, regeneration, and utilization, such as GLS1/GLS2 and GCLM/GCLC; (3) regulation of XCT to increase the intracellular cysteine pool in support of GSH synthesis; and (4) upregulation of UDP-glucuronosyltransferase (UGT) to promote glucuronidation pathway.

Furthermore, NRF2 was also found to regulate the expression of thioredoxin 1 (TXN1) and thioredoxin reductase 1 (TXNRD1), which can reduce oxidized protein thiols [25, 26]. TXN contains two redox-active cysteine residues and can be oxidized when reducing the disulfides within the oxidized proteins. The oxidized TXN can be reduced by NADPH-dependent TXNRD1. The TXN system is crucial for normal cellular function as TXN knockout mice were reported to be embryonic lethal [27]. TXN helps modulate the activity of many transcription factors through oxidoreductive modification of the protein. For example, a DNA repair protein, redox factor Ref-1, was reported to promote the DNA-binding activities of several transcription factors, including HIF-1α and p53, which confers cytoprotective function under hypoxia. The reducing environment promotes the DNA binding of HIF-1 in the nucleus, which is achieved by TXN-mediated Ref-1 reduction [28, 29].

Another important function of NRF2 in mediating antioxidant response is through promoting NADPH production. NRF2 promotes the expression of NADPH-generating enzymes, including the rate-limiting enzyme glucose-6-phosphate dehydrogenase (G6PD) in the pentose phosphate pathway (PPP); the isocitrate dehydrogenase (IDH), which converts isocitrate to α-ketoglutarate; and the malic enzyme 1 (ME1) that converts pyruvate to malate [30]. NADPH is a major source of reducing power in cells, which can be used to regenerate glutathione and thioredoxin. Additionally, other NRF2-regulated enzymatic processes, such as quinone detoxification by NQO1 and aldo-keto reductase (AKR), also require NADPH.

Glucuronidation is another process involved in xenobiotic metabolism, which involves conjugation of glucuronic acid, derived from UDP-glucuronic acid to a substrate, which is mediated by UDP-glucuronosyltransferase (UGT) enzymes. It is considered to be the major way to eliminate most of the xenobiotics. For example, the end product of heme metabolism, bilirubin, is eliminated through this pathway, which is catalyzed by the enzyme UDP-glucuronosyltransferase 1A1 (UGT1A1). In addition, various types of amino acids, such as taurine and glycine, can also be attached to the molecules to assist excretion. Furthermore, it has been found that S-adenosyl-L-methionine (SAM) can provide a methyl group for the conjugation reaction. For example, arsenic, which is known to generate oxygen-based radicals, can be detoxified through the reaction mediated by As(III) S-adenosylmethionine methyl-transferases [31]. Another kind of detoxification pathway, called sulfation or sulfonation, is catalyzed by a group of enzymes, termed sulfotransferases (SULTs), which are responsible for transferring sulfuryl group donated by 3′-phosphoadenosine-5′-phosphosulfate (PAPS) to the hydroxyl or amine groups [32]. The major function of sulfonation was known to modulate the receptor activity of estrogen and androgen and steroid biosynthesis [33, 34]. The inorganic sulfate used for the phase II sulfation pathway is derived from a process called sulfoxidation, which converts those sulfur amino acids (e.g., cysteine and methionine) into the sulfate.

Quinones are chemicals that contain quinoid ring, which can be converted to semiquinones by NADPH: cytochrome P450 reductase. Semiquinones react with oxygen and generate ROS, which leads to oxidative damage to cells. To prevent ROS production from semiquinones, the cytosolic, NRF2-regulated flavoproteins NAD(P)H: quinone oxidoreductases (NQOs) and aldo-keto reductase (AKR) can compete with P450 reductases and convert quinones to a relatively stable hydroquinones (quinols) [35]. In addition, NRF2 also regulates the expression

http://dx.doi.org/10.5772/65141

of ferritin subunits to promote free Fe^{2+} sequestration, which can inhibit Fe^{2+}-dependent free radical generation, and thus reducing the Fe^{2+}-mediated toxicity [36].

3. The regulation of cellular metabolism by reactive oxygen species

Reactive oxygen species have pleiotropic effects on cellular processes, including the modulation of cellular signaling and metabolism (**Figure 3**). Following oxidative insult, cellular metabolism is reprogrammed to increase the production of cellular building blocks to replace damaged cellular components. Changes in the cellular redox state can alter the oxidation state of metabolic enzymes, resulting in changes in their activity and altered flux through upstream and downstream pathways. Glyceraldehyde-3-phosphate dehydrogenase (GAPDH) is one of the most oxidized proteins in response to hydrogen peroxide [37]. It has been proposed that this serves to reroute metabolic intermediates into the pentose phosphate pathway to facilitate NADPH generation and adaptation to oxidative stress [38]. Indeed, GAPDH mutants that cannot be oxidized impair the cell's ability to generate NADPH following oxidative insult [39]. Curiously, oxidized GAPDH interacts with RNA- and DNA-binding proteins, suggesting that these adaptation mechanisms may not be limited to alterations in enzyme activity [40]. Glycolytic intermediates also build up further downstream in response to oxidative insults. Pyruvate kinase M2 (PKM2) is also inhibited by oxidation [41], which enhances flux through the pentose phosphate pathway and oxidative stress resistance. PKM2 also controls flux into the serine biosynthesis pathway [42], which supports glutathione production [43, 44]. The glycolysis modulator TIGAR, a fructose-2,6-bisphosphatase (Fru-2,6-BP), also plays a key role in the modulation of metabolism in response to oxidative stress. TIGAR is induced in response to ROS [45] and protects cells from ROS-induced cell death [46]. TIGAR modulates the intracellular levels of fructose-2,6-bisphosphate (Fru-2,6-BP), thereby regulating the activity of PFK1 and flux through glycolysis. PFK1 inhibition diverts upstream metabolites into the PPP, thereby increasing production of NADPH to support the antioxidant response [46–49]. TIGAR also inhibits ROS by binding to hexokinase 2 (HK2) and promoting its activity at the mitochondria during hypoxia [50]. This binding is independent of the bisphosphatase activity of TIGAR, suggesting that the regulation of the PPP by TIGAR is both phosphatase dependent and independent. Thus, modulation of glycolytic enzyme activity by ROS supports the antioxidant response.

Oxidation of enzymes in the TCA cycle, electron transport chain (ETC), and β-oxidation pathway also occurs following oxidative stress. Oxidation can either be reversible or irreversible, depending on the degree of oxidative stress. Reversible oxidation is temporary and occurs through disulfide bond formation, S-glutathionylation, or oxidation of cysteine residues to sulfenic acid. Irreversible oxidation occurs through oxidation of cysteine residues to sulfinic or sulfonic acid. Temporary S-glutathionylation of enzymes in the TCA cycle, ETC, or β-oxidation pathway protects them from irreversible oxidation when ROS is high but may also hypothetically serve as a mechanism to diminish mitochondrial ROS production by reducing their activity [51].

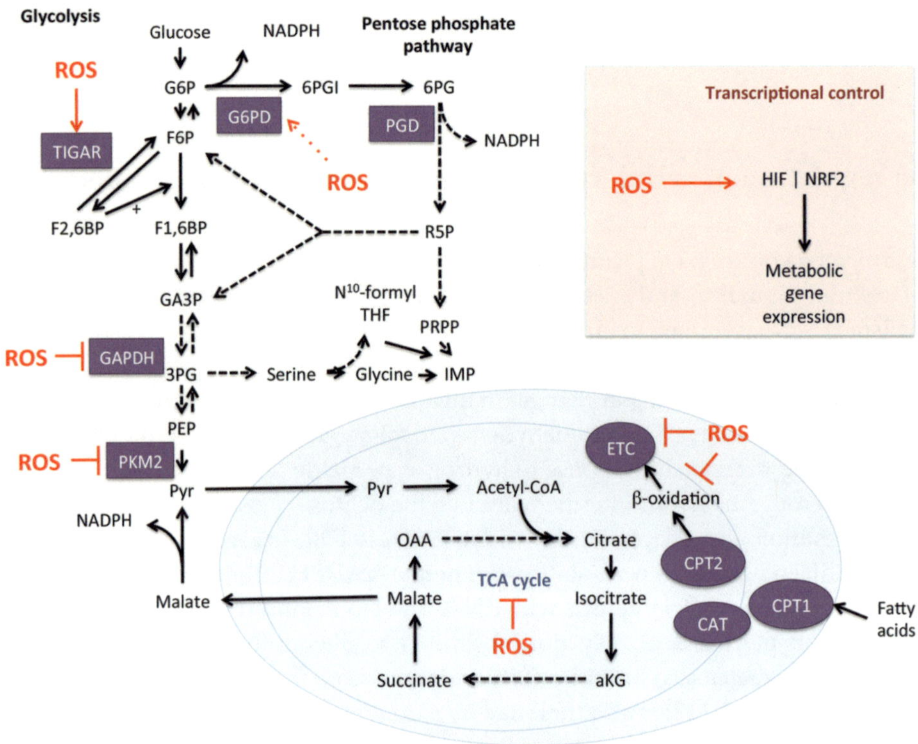

Figure 3. Modulation of cellular metabolism by ROS. ROS have pleiotropic effects on cellular metabolism, including the inhibition of the glycolytic enzymes glyceraldehyde-3-phosphate dehydrogenase (GAPDH) and pyruvate kinase M2 (PKM2), which results in accumulation of upper glycolytic intermediates and increased flux through the pentose phosphate pathway (PPP). ROS also promote PPP flux through the activation of TIGAR and by lowering cellular NADPH levels, thereby increasing the activity of glucose-6-phosphate dehydrogenase (G6PD). ROS inhibit reactions in the mitochondria, including the TCA and β-oxidation, which may serve to lower the production of ROS in already stressed cells. Furthermore, ROS induce transcriptional changes that lead to metabolic alterations through the activation of the transcription factors HIF and NRF2.

Reactive oxygen species also affect the activity of transcription factors that control metabolic enzyme expression. Two of the best-known examples of ROS-regulated transcription factors are NRF2 and hypoxia-inducible factors HIF1α and HIF2α. There are many parallels between these transcription factors. First, their regulation is very similar. Under basal, unstressed conditions, their levels are kept low due to targeted degradation by the proteasome. For HIF transcription factors, this occurs by hydroxylation at conserved proline residues by prolyl hydroxylases under normoxia, allowing their recognition and ubiquitination by the VHL E3 ubiquitin ligase, which targets HIFs for proteasomal degradation. Under hypoxia, the activity of the prolyl hydroxylases is inhibited, due to both lower oxygen availability and ROS generation by the mitochondria. ROS induce oxidative dimerization of the HIF prolyl hydroxylase, leading to its inactivation and HIF stabilization [52]. HIF activation reprograms cellular metabolism by upregulating glycolytic gene expression in inhibiting pyruvate entry into the TCA cycle [53], thereby mediating a shift from oxidative phosphorylation to glycolysis. This metabolic reprogramming has been shown to suppress ROS production in cells [54].

http://dx.doi.org/10.5772/65141

4. NRF2 reprograms cellular metabolism to funnel metabolic intermediates into antioxidant pathways

While NRF2 regulates the expression of antioxidant and NADPH-generating enzymes, metabolic intermediates are required to support these pathways. Recent studies have demonstrated that NRF2 also regulates the expression of metabolic enzymes that funnel intermediates into pathways that support antioxidant function (**Figure 4**). In this section, we will discuss how NRF2 reprograms cellular metabolism to support cellular detoxification and repair.

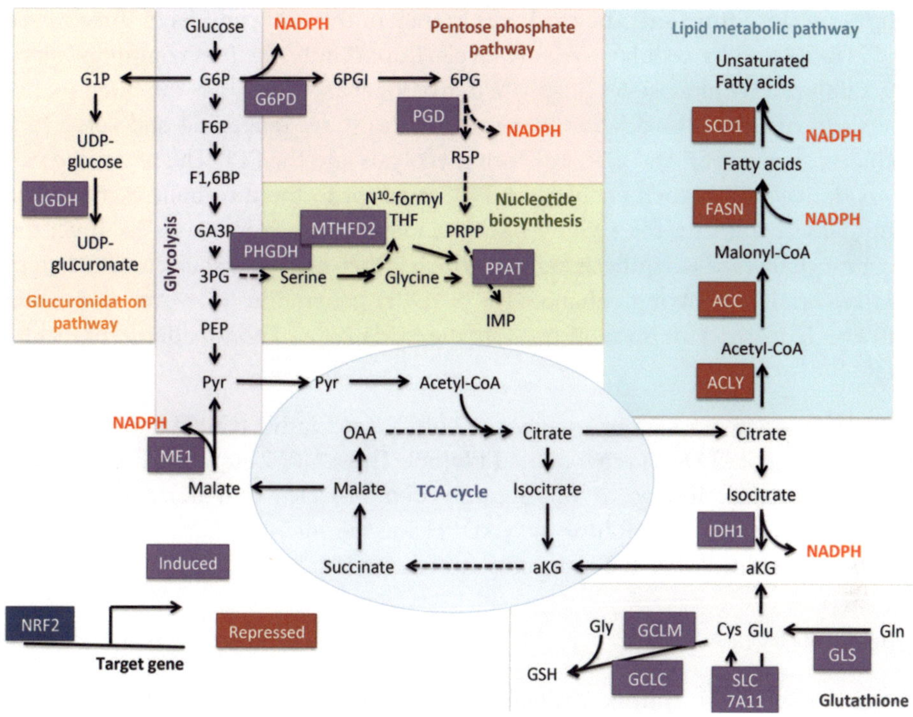

Figure 4. NRF2 reprograms cellular metabolism to support the antioxidant response. NRF2 induces many enzymes (denoted in purple) to increase the flux through metabolic pathways that support the antioxidant response, including the PPP (G6PD, PGD), nucleotide biosynthesis pathway (PHGDH, MTHFD2, PPAT), glucuronidation pathway (UGDH), enzymes that metabolize TCA cycle intermediates (ME1, IDH1), and enzymes in the glutathione biosynthesis pathway (GLS, SLC7A11, GCLM, GCLC). NRF2 also concomitantly represses lipid biosynthesis (ACLY, ACC, FASN, SCD1) to spare NADPH for detoxification reactions. The alterations in cellular metabolism serve to enhance the NRF2-induced transcriptional upregulation of metabolic enzymes by providing substrates for the enzymatic reactions.

4.1. Glucose metabolism

Many NRF2-regulated detoxification pathways rely on a supply of glucose for their activity. Thus, it is logical that NRF2 would not only regulate the enzymes in those pathways but the supply of metabolites that feed into them. Indeed, NRF2 activation leads to increased glucose uptake and glucose addition in fibroblasts [55]. Furthermore, interference with the supply of glucose was found to inhibit NRF2-mediated detoxification of reactive species. In this section, we will discuss NRF2-regulated pathways that utilize glucose.

4.1.1. The pentose phosphate pathway

Following phosphorylation by hexokinase, glucose enters glycolysis as glucose-6-phosphate (G6P). At this step, G6P can either continue through glycolysis or be diverted into pentose phosphate pathway, which consists of the oxidative and non-oxidative branches [56]. The oxidative branch produces both ribose-5-phosphate (R5P), which is precursor for nucleotide synthesis, and NADPH, which is not only required for biosynthesis reactions but also for antioxidant response function. The oxidative branch of the PPP consists of three irreversible steps and is regulated by cellular NADPH levels. Importantly, the first committed step of the pathway, catalyzed by glucose-6-phosphate dehydrogenase (G6PD), is inhibited by NADPH [57]. The steps of the non-oxidative branch of the PPP are reversible and serve to funnel intermediates, such as F6P and G3P, between glycolysis and the PPP. The reversible nature of the non-oxidative PPP branch enables the PPP to adapt to the metabolic demands of cells. Following oxidative insult, the activity of the oxidative branch increases to direct the non-oxidative branch toward resynthesizing F6P, which is then converted back to G6P to replenish the oxidative branch, allowing enhances for NADPH generation. However, in proliferating cells with a high demand for nucleotides, both the oxidative and non-oxidative branches serve to generate R5P.

NRF2 regulates the expression of enzymes in both the oxidative (G6PD and PGD) and non-oxidative (TKT and TALDO1) arms of the PPP [58]. Thus, NRF2 supports both NADPH and nucleotide production. Indeed, it was recently demonstrated that the pentose phosphate pathway (PPP), a major contributor of NADPH for the maintenance of glutathione in its reduced state, is critical for NRF2-induced proliferation [58, 59].

4.1.2. Nucleotide biosynthesis

The de novo synthesis of purine nucleotides proceeds from R5P in a 10-step pathway to produce inosine monophosphate (IMP), which is subsequently metabolized to AMP and GMP. The first step is the synthesis of phosphoribosyl pyrophosphate (PRPP), in which the enzyme PRPS1 transfers a pyrophosphate group onto R5P. Next, PPAT, a NRF2-regulated enzyme [59], catalyzes the displacement of the pyrophosphate with an amide nitrogen from glutamine. The subsequent steps involve the incorporation of glycine, 2 N_{10}-formyl-THF units, aspartate, and glutamine-derived amide groups. In contrast, the synthesis of pyrimidines starts from ammonia and bicarbonate and proceeds in a six-step pathway that involves in the incorporation of aspartate and PRPP. By regulating the activity of the PPP and the levels of R5P, NRF2 may influence the production of both purines and pyrimidines. NRF2 also influences cellular levels of glycine, N_{10}-formyl-THF, and glutamine, which will be discussed in the following sections. Thus, NRF2 supports nucleotide biosynthesis through direct transcriptional mechanisms and by rewiring cellular metabolism to funnel metabolic intermediates into the pathways.

4.1.3. The serine biosynthesis pathway

The amino acids, serine and glycine, are required for many biological processes. Beyond protein synthesis, serine is essential for the synthesis of sphingolipids and most phosphatidylserine head groups, an important component of cellular membranes [60, 61]. Glycine contributes one nitrogen and two carbon atoms to the purine ring for nucleotide biosynthesis. Serine and glycine also charge the folate pool with one-carbon units, including N_{10}-formyl-THF, that are ultimately used to produce nucleotides, methionine, and S-adenosylmethionine, the latter of which is the major methyl donor for methyltransferase reactions including those of DNA and histones [62, 63]. Serine is also required for the transsulfuration pathway that generates cysteine. Furthermore, both cysteine and glycine are incorporated into the tripeptide glutathione, and thus serine is critical for both cellular redox homeostasis and xenobiotic metabolism processes [64, 65]. Consequently, intermediary metabolism is organized so that serine and glycine are positioned to play essential roles in biomass accumulation, DNA replication, epigenetics, and redox homeostasis.

Cells obtain serine and glycine through import from the extracellular space or de novo synthesis. Amino acids are transported by systems which are broadly defined based on their physiochemical properties and substrate specificity [66]. Multiple systems import serine, including the commonly expressed ASC system that mediates the symport of alanine, serine, or cysteine with sodium [67]. De novo synthesis of serine from the glycolytic intermediate 3-phosphoglycerate (3-PG) occurs through a pathway consisting of three sequential steps: phosphoglycerate dehydrogenase (PHGDH) oxidizes 3-PG to 3-phosphohydroxypyruvate (pPYR) using an NAD^+ cofactor, phosphoserine aminotransferase (PSAT) transaminates pPYR to phosphoserine (pSER) using glutamate as the nitrogen donor, and finally phosphoserine phosphatase (PSPH) dephosphorylates pSER to produce serine [68]. Importantly, the serine thus produced can be converted to glycine via serine hydroxymethyltransferases concomitantly charging the folate pool with one-carbon units.

Recently, we found that NRF2 regulates the serine biosynthesis pathway through ATF4 and PHGDH, which led to an independence from exogenous serine sources [43]. We observed that serine-derived glycine was found to support glutathione metabolism. While most of the cysteine for glutathione was derived from exogenous source in nutrient replete culture, serine-derived cysteine may become more important under nutrient-limiting conditions. Additionally, glycine and intermediates of the folate cycle were found to support the biosynthesis of both purines and pyrimidines, which synergized with the regulation of the pentose phosphate pathway to supply ribose for nucleotides. By regulating both the production of serine and glycine, as well as the enzymes that utilize these amino acids, NRF2 may better influence their fate by funneling them into antioxidant response pathways.

The serine biosynthesis pathway may support the NRF2-regulated antioxidant program in multiple ways. First, the ability to supply glutathione through the availability of the metabolite substrates for de novo GSH synthesis supports ROS detoxification. Furthermore, the production of nucleotides supports the replenishment of the oxidized nucleotide pool and facilitates DNA repair. Additionally, serine-derived metabolites such as glutathione, glycine, cysteine, and SAM support phase II metabolism because they are utilized for conjugation with toxins for toxin metabolism and excretion.

4.1.4. Glucuronidation pathway

In the process of glucuronidation, UDP-glucuronosyltransferases, such as the NRF2-regulated enzyme UGT1A1 [69], transfer the glucuronic acid component of UDP-glucuronic acid onto substrates, including drugs, bilirubin, hormones, and other molecules. This serves to increase the solubility of these molecules for their excretion. UDP-glucuronic acid is generated from G6P in a multistep process. First, phosphoglucomutase converts G6P to glucose-1-phosphate (G1P). Next, the UDP molecule is attached by UDP-glucose pyrophosphorylase to generate UDP-glucose. Finally, UDP-glucuronate is generated from UDP-glucose by UDP-glucose dehydrogenase, a NRF2-regulated enzyme [70, 71]. Thus, NRF2 promotes the expression of enzymes that synthesize UDP-glucuronate, as well as glucose uptake to promote entry of G6P into this pathway.

4.1.5. The TCA cycle

Following pyruvate entry into the mitochondria, pyruvate is metabolized in the TCA cycle to generate NADH and FADH2 to fuel the electron transport chain but also metabolic intermediates for biosynthesis and NADPH production. Following the generation of acetyl-CoA by pyruvate dehydrogenase (PDH), acetyl-CoA condenses with oxaloacetate to form citrate. Citrate can be exported to the cytoplasm, where it is used as precursor for lipid biosynthesis or metabolized in the mitochondria to isocitrate by aconitase. In the next step, isocitrate dehydrogenase (IDH) metabolizes isocitrate to α-ketoglutarate (α-KG), which is further metabolized to succinyl-CoA by α-KG dehydrogenase complex. Succinyl-CoA is then transformed to succinate by the succinyl-CoA synthetase. Succinyl-CoA is oxidized to fumarate by the succinate dehydrogenase (SDH) complex and then hydrated to malate by fumarate hydratase (FH). Oxidation of malate, catalyzed by malate dehydrogenase, finally regenerates oxaloacetate, and the cycle continues.

There are several points of the TCA cycle that are critical for NRF2-regulated processes. First, NRF2 promotes PDH flux to increase substrate entry into the TCA cycle and also promotes cellular glucose oxidation [58]. This results in increased TCA cycle activity and sufficient substrate availability for NRF2-regulated enzymes that require TCA cycle intermediates. These enzymes include malic enzyme 1 (ME1) and IDH1, which generate NADPH, and thus the levels and subcellular compartmentalization of malate and isocitrate dictate the activity of these enzymes. ME1 is a mitochondrial enzyme that metabolizes malate to pyruvate to generate NADPH. IDH1 is a cytosolic enzyme that metabolizes isocitrate to α-KG to generate cytosolic NADPH. Furthermore, pyruvate carboxylase was identified as a potential NRF2-MafG target gene [72]. Pyruvate carboxylase catalyzes the carboxylation of pyruvate to oxaloacetate. In cells that have significant malic enzyme activity, this enzyme could play a critical anaplerotic role by replenishing oxaloacetate so the TCA cycle can progress.

4.2. Glutamine metabolism

Glutamine is the most abundant free amino acid in human serum, with average concentrations of around 500 μM. Proliferating cells in culture are highly dependent on glutamine, as it is both a key nitrogen donor and carbon supply for the TCA cycle. NRF2 controls the uptake, metabolism, and fate of glutamine in cells in addition to its role in regulating glucose metabolism. As mentioned previously, NRF2 regulates the expression of ATF4, which controls the expression of the glutamine transporter SLC1A5 [73], and asparagine synthetase (ASNS), which generates both asparagine and glutamate from aspartate and glutamine. Indeed, we found that NRF2 controls the binding of ATF4 to the ASNS promoter [43]. Additionally, glutaminase (GLS), which metabolizes glutamine to glutamate, is a NRF2 target gene [74]. Here, we will discuss NRF2-regulated pathways that utilize glutamine and their role in supporting the antioxidant response.

4.2.1. Glutathione biosynthesis

As glutamate is one of the three amino acid constituents of glutathione and can also supply the TCA cycle through the production of α-KG, NRF2 may support both the antioxidant response and NADPH production through the control of glutamine metabolism. Indeed, when Nrf2 is inhibited, the carbon flux from glutamine into GSH biosynthesis is severely diminished [59]. Furthermore, glutamate is an obligate exchange molecule for the NRF2-regulated cysteine transporter SLC7A11, an obligate antiporter. Cysteine is also an amino acid constituent of glutathione, and glutamate production may support glutathione metabolism by increasing the intracellular availability of cysteine. Glutamine and glutamate are also the nitrogen donors for many key metabolic pathways. Glutamate donates the nitrogen for the transamination step in serine biosynthesis, which can supply both glycine and cysteine for glutathione production.

4.2.2. Other biosynthesis

As described earlier, glutamine is required for two steps in purine biosynthesis, whereas aspartate, which acquires its nitrogen from glutamate, is required for a step in purine biosynthesis and a step in pyrimidine biosynthesis. Additionally, glutamine is a major source of TCA cycle carbon. Glutamine-derived α-KG enters the TCA cycle and in many cell lines contributes more carbon to TCA cycle intermediates than glucose. Thus, glutamine entry into the TCA cycle can provide metabolic intermediates for NRF2-regulated NADPH-producing enzymes, such as IDH1 and ME1.

4.3. Lipid metabolism

The effects of NRF2 on lipid metabolism are less well characterized. Studies have found that increased NRF2 signaling in the mouse liver is associated with repression of lipogenesis [75]. Suppression of lipogenesis plays an important role in the NRF2-regulated antioxidant response. While NRF2 activates NADPH production through the upregulation of the pentose phosphate pathway, malic enzyme, and isocitrate dehydrogenase, NRF2 also supports cellular

NADPH levels by suppressing NADPH-consuming processes. Lipid biosynthesis is one of the most NADPH-consuming processes in the cell. Wu et al. found that NRF2 suppresses both lipid biosynthesis and desaturation genes in the mouse liver, including NADPH-utilizing enzymes fatty acid synthase (FASN) and stearoyl-CoA desaturase 1 (SCD1) [30]. Additionally, high-fat diet-induced lipid biosynthesis enzyme induction, including *Acly*, *Acaca*, and *Fasn*, is more pronounced in the livers of $Nrf2^{-/-}$ mice [76]. Suppression of lipogenesis thereby increases the availability of NADPH for use by the antioxidant response.

Activation of NRF2 also promotes fatty acid oxidation. NRF2 knockdown was found to suppress the expression of fatty acid oxidation genes ACOX1, ACOX2, CPT1, and CPT2 [77], and NRF2 activation in the mouse lung induces the expression of fatty acid oxidation genes and lipases. Activation of fatty acid oxidation may serve several functions. First, it may induce the degradation of damaged lipid molecules so they are not utilized for membrane synthesis. Second, fatty acid oxidation has been shown to provide NADPH for ROS detoxification [78].

4.4. Heme and iron metabolism

Another critical function of the NRF2 antioxidant program is the regulation of iron metabolism. NRF2 regulates heme degradation, thus removing excess amounts of the prooxidant molecule heme from the cellular pool. Heme oxygenase metabolizes heme to iron and biliverdin. Free iron is reactive and dangerous to cells, and NRF2 regulates the expression of ferritin heavy and light chains (FTL and FTH1), which immediately bind free iron, as well as ferroportin, which exports it [79]. The NRF2 target biliverdin reductase metabolizes biliverdin to bilirubin [30]. In the body, bilirubin is excreted as waste, but bilirubin can serve as an antioxidant molecule, and the bilirubin-biliverdin redox couple has direct antioxidant function [80]. Furthermore, bilirubin inhibits NADPH oxidase, which serves to keep ROS levels low.

Although it may seem counterintuitive, NRF2 also regulates enzymes involved in heme biosynthesis. Ferrochelatase (FECH) is a direct NRF2 target gene [36] that is responsible for the final step in heme biosynthesis. Importantly, the first step in heme biosynthesis requires the amino acid glycine, the production of which is NRF2 regulated. While the heme may have prooxidant functions in large quantities, heme is constantly turned over and excreted from cells as bilirubin [81]. The synthesis of biliverdin and bilirubin actually protects cells from oxidative stress, possibly due to the antioxidant functions described above.

4.5. Autophagy and the proteasome

Large intracellular particles, including damaged mitochondria and protein aggregates, are degraded by a process known as autophagy. Intracellular components are first engulfed in double membrane vesicles called autophagosomes, which subsequently fuse with lysosomes for digestion of their contents and recycling of their building blocks. The autophagy adapter molecule p62 is a direct target of NRF2 [82]. While p62 may play a role in the stress-responsive degradation of damaged intracellular components, it remains to be determined whether autophagy is critical for the NRF2-regulated antioxidant response. Rather, the relationship between p62 and NRF2 is much more complex. Studies have identified that autophagy deficiency leads to NRF2 activation due to p62 accumulation [83]. p62 was found to bind to the NRF2-binding site on KEAP1, thereby competing with NRF2 for KEAP1 binding [84]. However, not only does p62 prevent NRF2 binding to KEAP1, it induces autophagic degradation of KEAP1, resulting in a positive feedback loop of NRF2 activation [82]. This process is regulated, as mTORC1-dependent phosphorylation of p62 increases its affinity for KEAP1, thereby activating NRF2 [85]. Furthermore, there may be other components of this process. Sestrins, which have long been linked to protection from oxidative stress, were found to interact with p62 and KEAP1 to promote KEAP1 autophagic degradation [86]. Sestrin 2 is a leucine sensor that inhibits mTORC1 signaling [87], providing an interesting link between amino acid sensing, mTOR, and NRF2-KEAP1 pathway.

NRF2 also regulates the removal of damaged and misfolded proteins through the regulation of the proteasome. Unwanted proteins are degraded by the 26S proteasome, which consists of a 20S core and a 19S regulatory subunit. NRF2 regulates the 20S proteasome subunits PSMA1, PSMA7, PSMB3, PSMB5, and PSMB6 [88–90], as well as the proteasome activator PA28αβ [73]. Importantly, induction of the proteasome is required for adaptation to oxidative stress [91] and can extend the life span of human fibroblasts [92].

5. Consequences for tissue homeostasis and disease

In the previous section, we have discussed how NRF2-regulated cellular metabolism supports the antioxidant program. In this section, we will focus on how these programs contribute to both normal physiology and disease.

5.1. NRF2 and tissue homeostasis

In normal cells exposed to oxidative insults, these pathways work in concert to produce building blocks for the replacement and repair of damaged components. Recently, several studies demonstrated that the NRF2-regulated antioxidant response is crucial for tissue homeostasis. The recent work by Telorack et al. unraveled the function of GSH in keratinocytes and showed that GSH deficiency affects keratinocyte survival and wound repair. By using mice deficient in glutamate-cysteine ligase (GCL) in keratinocytes, the authors found that GSH deficiency results in more DNA damage and cell death, suggesting a crucial role for the antioxidant capacity in skin integrity maintenance [93]. In addition, NRF2 was reported to confer radioprotection to human lung fibroblasts through upregulating miR-140 expression, which is involved in the regulation of lung fibroblast self-renewal [94]. In the following sections, we will discuss how the regulation of metabolism and ROS levels by NRF2 contributes to tumor development, aging, and stem cell function.

5.2. NRF2 and cancer

NRF2 is frequently mutated in many types of cancers, with the mutations clustered around the DLG (43 %) and ETGE (57 %) motifs, which are the two critical KEAP1 interaction domains [95]. Furthermore, mutations in *KEAP1* that abolish KEAP1 function were also found to activate

NRF2 and promote the growth of lung cancer cells [96]. A recent study characterizing the genomic alterations in squamous cell lung cancers has found the NRF2-KEAP1 pathway that altered in 34% of the 178 tumor samples examined [97]. Constitutive activation of NRF2 has been considered as a resistance mechanism to chemo or radiation therapy because the NRF2-mediated antioxidant response eliminates therapy-induced ROS, thereby promoting cell survival. For example, NRF2 activation was found to protect cells against ionizing radiation toxicity and confer radioresistance [98].

In addition, aberrant activation of NRF2 also serves to increase the production of building blocks to sustain uncontrolled proliferation in cancer. The activation of the serine biosynthetic pathway may represent a significant contribution to the pro-tumorigenic activity of NRF2. We and others have observed that enhanced de novo synthesis of serine and glycine from glucose reduces the reliance of cells on extracellular sources [43, 99]. Additionally, many NRF2-regulated metabolic pathways were found to confer growth advantages to cancer cells, such as the upregulation of the PPP, nucleotide synthesis, and NADPH production. Mitsuishi et al. have shown that NRF2 promotes nucleotide production via upregulation of PPP and glutamine utilization in response to proliferating signals, which in turn accelerates tumor growth [59]. Interestingly, another study showed that NRF2 overexpression increased human telomerase reverse transcriptase (hTERT) levels and upregulated the expression of two PPP enzymes, G6PD and TKT, which could promote the progression of glioblastoma [100]. Moreover, NRF2-regulated TKT expression has also been found to promote cancer development via counteracting oxidative stress [101].

Furthermore, the NADPH-producing metabolic enzyme ME1 which links glycolysis with the TCA cycle was reported to promote the metastatic potential of HCC and correlated with poor prognosis [102]. Another study has observed that ME1 could promote tumor growth, potentially through increasing NADPH production, which is required for GSH regeneration [103]. Additionally, another NADPH-generating enzyme IDH1 was found to be crucial for the anchorage-independent growth, a property shared by many malignant cells. It was shown that cell detachment promotes ROS production in the mitochondria. The oxidative stress in mitochondria could be counteracted by IDH1-driven reductive carboxylation in the cytosol, followed by the transfer of citrate into the mitochondria, which was to be oxidized by IDH2 to generate NADPH [104]. Moreover, research published by Saito et al. has found that p62-mediated NRF2 activation in HCC could promote tumor malignancy via redirecting glucose into the glucuronate pathway and glutamine toward GSH synthesis [105]. These studies demonstrate that multiple NRF2-regulated metabolic enzymes play important roles in cancer.

5.3. NRF2 and aging

Oxidative stress has been suggested to cause several pathologies, and many of which are age related. The oxidative stress theory of aging came early in 1956 from Denham Harman, who proposed that aging is attributed to the side effects of free radicals on cell constituents and that mitochondria are the key organelles associated with aging [106]. This is supported by the finding that aging leads to more oxidation of the GSH and more damaged mtDNA in the mitochondria, which can be prevented by oral antioxidant administration. In fact, the levels of oxo-8-deoxyguanosine, which is considered to be the major oxidative damage to DNA, have been found correlated linearly with oxidation of glutathione [107]. Moreover, aging was associated with the decline of antioxidant response, as some of the antioxidant genes induced by oxidative stress declined with age, such as GSH levels [108]. However, some of the changes of the antioxidant genes in the aged mice are inconsistent, and the phenotype might be tissue specific. Overall, the age-dependent changes of the antioxidant genes suggest that ROS levels might contribute in part for the aging processes. As mentioned in the previous section, NADPH production is one of the important antioxidant responses to protect the cell from ROS-induced damage. Interestingly, the key enzyme G6PD, which catalyzes the rate-limiting step of the PPP has been found to extend the life span of transgenic *Drosophila melanogaster* [109]. A similar finding was observed in mice, in which G6PD overexpression could protect the transgenic mice from oxidative damage and improve their life span through the elevation of NADPH levels, thereby lowering the level of ROS-induced damage (8-OHdG) [110].

Because increased oxidative stress is thought to be a major characteristic feature of aging and NRF2 has long been considered as the major regulator of the cellular antioxidant response, the potential role of NRF2 in aging is intriguing. Indeed, there have been several studies connecting NRF2 with aging. The study from Hirota et al. showed that $Nrf2^{-/-}$ mice accelerated UVB-induced photoaging, including the phenotypes of loss of skin flexibility and wrinkle formation [111]. Additionally, another group observed that the hearing ability of NRF2-deficient mice was more impaired and the number of hair cells was significantly reduced compared to the wild-type mice [112]. Furthermore, progressive loss of NRF2 activity was seen with aging, which may be explained by studies showing that the expression of NRF2 positive regulators P62 and BRCA1 decreased with age, while NRF2 negative regulators, KEAP1 and BACH1, increased with age [113].

Another potential mechanism for age-dependent loss of antioxidant genes was provided by one of the studies showing that the presence of transcription repressor BACH1 along with the absence of coactivator CREB-binding protein (CBP) in the aged mice may convert NRF2 binding from an active ARE to an alternative ARE [114]. Interestingly, a recent paper published by Kubben et al. found that repression of NRF2 antioxidant pathways served as a driver mechanism for the genetic premature aging disease Hutchinson-Gilford progeria syndrome (HGPS) [115], which is caused by constitutive progerin production, a mutant version of the nuclear protein lamin A. Progerin was known responsible for many cellular defects, one of which is the attrition of mesenchymal stem cells (MSCs) that leads to tissue defect [116]. This study demonstrated that sequestration of NRF2 by progerin, which results in protein mislocalization and the impairment of NRF2-mediated antioxidant response, contributes to progerin-associated nuclear aging defects [115]. Additional studies are still required to further support the role of NRF2 in the aging processes.

5.4. NRF2 and stem cells

Regulation of ROS levels in cells is important, as excessive ROS can lead to cellular damage. In contrast, a modest level of ROS is crucial for maintaining proper biological functions. The redox status has been suggested to regulate stem cell self-renewal and differentiation, and accumulating evidence indicates that stem cells undergo self-renewal in the environment with low ROS levels, while ROS levels increase in differentiated stem cells [117, 118]. Because embryonic stem (ES) cells give rise to all the tissues in an organism, mechanisms are in place to avoid the accumulation of DNA to prevent mutations that would transmit to tissues and subsequent generations. To achieve this, ES cells have higher antioxidant defense mechanisms, resulting in a lower mutational frequency than differentiated cells [119, 120]. Furthermore, downregulation of ROS generation in stem cells is achieved through a reduced dependence on oxidative phosphorylation [121, 122]. This leads to a greater reliance on glycolysis, which serves to not only generate ATP but also metabolic intermediates to feed the pentose phosphate pathway. Thus, the metabolic reprogramming of stem cells to support antioxidant responses supports ROS detoxification and stem cell health.

As a redox sensor, NRF2 might also contribute to the regulation of stem cell function. Indeed, NRF2 was found constitutively active in intestinal stem cell (ISC) in *Drosophila*, and KEAP1-mediated NRF2 repression was required for ISC proliferation. The redox balance controlled by NRF2-KEAP1 is critical for intestinal homeostasis, as evidenced by NRF2 loss leads to age-related degeneration of the intestinal epithelium [123]. Moreover, NRF2 is required for the survival of hematopoietic stem progenitor cells (HSPCs) and the development of myeloid cells in mice, as NRF2 KO bone marrow had defect in stem cell function and showed reduction of chimerism after transplantation [124]. A recent study has found that the deacetylase SIRT6 maintained the homeostasis of human mesenchymal stem cells (hMSCs) via regulation of redox metabolism through coactivating NRF2. SIRT6-null hMSCs showed a premature cellular attrition and could be rescued by overexpressing HO-1, suggesting that an imbalance of ROS levels could result in stem cell decay [125]. Another recent published work demonstrated that reduced NRF2 expression is responsible for the decline of neural stem/progenitor cell (NSPC) function, as it is found that NRF2 expression was suppressed in the aged NSPC, and overexpression of NRF2 in those old cells could render them to a similar state as the young NSPCs, by showing increasing cell survival and regeneration [126].

In addition, NRF2 activation was found crucial for the reprogramming of induced pluripotent stem cell (iPSC), which was known to be an inefficient process and required the metabolic shift from the oxidative phosphorylation to glycolysis as a major way to generate ATP. It is shown that at the early phase of iPSC reprogramming, ROS levels increase, which activate NRF2, and it is important for navigating the transition [127]. Additionally, high glutathione and glutathione peroxidase 2 levels were found critical for maintaining genomic integrity of human pluripotent stem cells, which were known highly sensitive to ROS-induced cell death [128].

Besides modulating the function of normal stem cell, it is reported that redox balance could also affect cancer stem cell (CSC) function. CSCs or tumor-initiating cells are small populations of cells with stem-like properties, which were considered responsible for drug resistance and tumor relapse. A study has found that NRF2 is responsible for maintaining the self-renewal function of glioma stem cells (GSCs), supported by the evidence that NRF2 deficiency attenuated the tumorigenicity of GSCs in the xenograft model [129]. Interestingly, another study using the lung cancer model observed a downregulated expression of 26S proteasome in lung cancer stem cells. As 26S was known to target NRF2 for degradation, it is postulated that reducing NRF2 degradation might result in high levels of NRF2 expression which confers the growth advantage to CSCs [130].

6. Conclusions

In this chapter, we have discussed how NRF2 controls the antioxidant response and reprograms cellular metabolism to support antioxidant function, as well as the role of these pathways in tumorigenesis, stem cell function, and aging. Taken together, these studies demonstrate that cellular modulation of ROS levels is highly important for maintaining the proper function of many cellular processes. Understanding how NRF2 coordinates different metabolic pathways to support cellular detoxification opens new avenues for therapeutic intervention. Proper tissue homeostasis occurs when the right amount of NRF2 activity is achieved. Too much NRF2 activity is associated with uncontrolled proliferation and neoplasia. Too little NRF2 activity results in a decline in stem cell function and aging. Restoring the balance could help treat these diseases. Because cancer cells hijack the protective capability of NRF2 to promote drug resistance and proliferation, specific NRF2 inhibitors would be desirable for the treatment of cancers that have constitutively active NRF2. In contrast, for those diseases caused by NRF2 loss of function or a defect in one of the NRF2-regulated pathways, reactivating an alternative pathway that could also contribute to ROS detoxification might be a way to resolve the problem. Intensive studies are still required to unravel the crosstalk between different NRF2-regulated metabolic pathways and diseases and enable the development of a better strategy for combating human diseases.

Author details

Ting-Yu Lin[1], Lewis C. Cantley[1] and Gina M. DeNicola[2*]

*Address all correspondence to: Gina.DeNicola@moffitt.org

1 Meyer Cancer Center, Weill Cornell Medicine, New York, NY, United States

2 Department of Cancer Imaging and Metabolism, Moffitt Cancer Center and Research Institute, Tampa, FL, United States

References

[1] Handy, D.E. and J. Loscalzo, *Redox regulation of mitochondrial function.* Antioxid Redox Signal, 2012. 16(11): p. 1323-67.

[2] Rhee, S.G., Cell signaling. H_2O_2, a necessary evil for cell signaling. Science, 2006. 312(5782): p. 1882-3.

[3] Gorrini, C., I.S. Harris, and T.W. Mak, *Modulation of oxidative stress as an anticancer strategy*. Nat Rev Drug Discov, 2013. 12(12): p. 931-47.

[4] Chandel, N.S., et al., *Mitochondrial reactive oxygen species trigger hypoxia-induced transcription*. Proc Natl Acad Sci U S A, 1998. 95(20): p. 11715-20.

[5] Kincaid, M.M. and A.A. Cooper, *ERADicate ER stress or die trying*. Antioxid Redox Signal, 2007. 9(12): p. 2373-87.

[6] Tanaka, H., et al., E2F1 and c-Myc potentiate apoptosis through inhibition of NF-kappaB activity that facilitates MnSOD-mediated ROS elimination. Mol Cell, 2002. 9(5): p. 1017-29.

[7] Irani, K., et al., Mitogenic signaling mediated by oxidants in Ras-transformed fibroblasts. Science, 1997. 275(5306): p. 1649-52.

[8] DeNicola, G.M., et al., Oncogene-induced Nrf2 transcription promotes ROS detoxification and tumorigenesis. Nature, 2011. 475(7354): p. 106-9.

[9] Wakabayashi, N., et al., Keap1-null mutation leads to postnatal lethality due to constitutive Nrf2 activation. Nat Genet, 2003. 35(3): p. 238-45.

[10] Kobayashi, A., et al., Oxidative stress sensor Keap1 functions as an adaptor for Cul3-based E3 ligase to regulate proteasomal degradation of Nrf2. Mol Cell Biol, 2004. 24(16): p. 7130-9.

[11] Itoh, K., et al., An Nrf2/small Maf heterodimer mediates the induction of phase II detoxifying enzyme genes through antioxidant response elements. Biochem Biophys Res Commun, 1997. 236(2): p. 313-22.

[12] Niture, S.K., A.K. Jain, and A.K. Jaiswal, Antioxidant-induced modification of INrf2 cysteine 151 and PKC-delta-mediated phosphorylation of Nrf2 serine 40 are both required for stabilization and nuclear translocation of Nrf2 and increased drug resistance. J Cell Sci, 2009. 122(Pt 24): p. 4452-64.

[13] Cullinan, S.B., et al., Nrf2 is a direct PERK substrate and effector of PERK-dependent cell survival. Mol Cell Biol, 2003. 23(20): p. 7198-209.

[14] Joo, M.S., et al., AMPK facilitates nuclear accumulation of Nrf2 by phosphorylating at serine 550. Mol Cell Biol, 2016. 36(14): p. 1931-42.

[15] Karapetian, R.N., et al., Nuclear oncoprotein prothymosin alpha is a partner of Keap1: implications for expression of oxidative stress-protecting genes. Mol Cell Biol, 2005. 25(3): p. 1089-99.

[16] Sun, Z., et al., Keap1 controls postinduction repression of the Nrf2-mediated antioxidant response by escorting nuclear export of Nrf2. Mol Cell Biol, 2007. 27(18): p. 6334-49.

[17] Tong, K.I., et al., Different electrostatic potentials define ETGE and DLG motifs as hinge and latch in oxidative stress response. Mol Cell Biol, 2007. 27(21): p. 7511-21.

[18] Chen, W., et al., Direct interaction between Nrf2 and p21(Cip1/WAF1) upregulates the Nrf2-mediated antioxidant response. Mol Cell, 2009. 34(6): p. 663-73.

[19] Gorrini, C., et al., BRCA1 interacts with Nrf2 to regulate antioxidant signaling and cell survival. J Exp Med, 2013. 210(8): p. 1529-44.

[20] Lau, A., et al., A noncanonical mechanism of Nrf2 activation by autophagy deficiency: direct interaction between Keap1 and p62. Mol Cell Biol, 2010. 30(13): p. 3275-85.

[21] Shen, G. and A.N. Kong, Nrf2 plays an important role in coordinated regulation of Phase II drug metabolism enzymes and Phase III drug transporters. Biopharm Drug Dispos, 2009. 30(7): p. 345-55.

[22] Higgins, L.G., et al., Transcription factor Nrf2 mediates an adaptive response to sulforaphane that protects fibroblasts in vitro against the cytotoxic effects of electrophiles, peroxides and redox-cycling agents. Toxicol Appl Pharmacol, 2009. 237(3): p. 267-80.

[23] Lewerenz, J., et al., The cystine/glutamate antiporter system x(c)(-) in health and disease: from molecular mechanisms to novel therapeutic opportunities. Antioxid Redox Signal, 2013. 18(5): p. 522-55.

[24] Dalle-Donne, I., et al., *S-glutathionylation in protein redox regulation*. Free Radic Biol Med, 2007. 43(6): p. 883-98.

[25] Hawkes, H.J., T.C. Karlenius, and K.F. Tonissen, Regulation of the human thioredoxin gene promoter and its key substrates: a study of functional and putative regulatory elements. Biochim Biophys Acta, 2014. 1840(1): p. 303-14.

[26] Malhotra, D., et al., Global mapping of binding sites for Nrf2 identifies novel targets in cell survival response through ChIP-Seq profiling and network analysis. Nucleic Acids Res, 2010. 38(17): p. 5718-34.

[27] Nonn, L., et al., The absence of mitochondrial thioredoxin 2 causes massive apoptosis, exencephaly, and early embryonic lethality in homozygous mice. Mol Cell Biol, 2003. 23(3): p. 916-22.

[28] Ema, M., et al., Molecular mechanisms of transcription activation by HLF and HIF1alpha in response to hypoxia: their stabilization and redox signal-induced interaction with CBP/p300. EMBO J, 1999. 18(7): p. 1905-14.

[29] Nishinaka, Y., et al., Redox control of cellular function by thioredoxin; a new therapeutic direction in host defence. Arch Immunol Ther Exp (Warsz), 2001. 49(4): p. 285-92.

[30] Wu, K.C., J.Y. Cui, and C.D. Klaassen, *Beneficial role of Nrf2 in regulating NADPH generation and consumption*. Toxicol Sci, 2011. 123(2): p. 590-600.

[31] Ajees, A.A. and B.P. Rosen, As(III) S-adenosylmethionine methyltransferases and other arsenic binding proteins. Geomicrobiol J, 2015. 32(7): p. 570-576.

[32] James, M.O. and S. Ambadapadi, *Interactions of cytosolic sulfotransferases with xenobiotics*. Drug Metab Rev, 2013. 45(4): p. 401-14.

[33] Xu, C., C.Y. Li, and A.N. Kong, *Induction of phase I, II and III drug metabolism/transport by xenobiotics*. Arch Pharm Res, 2005. 28(3): p. 249-68.

[34] Hodges, R.E. and D.M. Minich, Modulation of metabolic detoxification pathways using foods and food-derived components: a scientific review with clinical application. J Nutr Metab, 2015. 2015: p. 760689.

[35] Shultz, C.A., et al., Specificity of human aldo-keto reductases, NAD(P)H:quinone oxidoreductase, and carbonyl reductases to redox-cycle polycyclic aromatic hydrocarbon diones and 4-hydroxyequilenin-o-quinone. Chem Res Toxicol, 2011. 24(12): p. 2153-66.

[36] Chorley, B.N., et al., Identification of novel NRF2-regulated genes by ChIP-Seq: influence on retinoid X receptor alpha. Nucleic Acids Res, 2012. 40(15): p. 7416-29.

[37] Baty, J.W., M.B. Hampton, and C.C. Winterbourn, *Proteomic detection of hydrogen peroxide-sensitive thiol proteins in Jurkat cells*. Biochem J, 2005. 389(Pt 3): p. 785-95.

[38] Ralser, M., et al., Dynamic rerouting of the carbohydrate flux is key to counteracting oxidative stress. J Biol, 2007. 6(4): p. 10.

[39] Peralta, D., et al., A proton relay enhances H_2O_2 sensitivity of GAPDH to facilitate metabolic adaptation. Nat Chem Biol, 2015. 11(2): p. 156-63.

[40] Hwang, N.R., et al., Oxidative modifications of glyceraldehyde-3-phosphate dehydrogenase play a key role in its multiple cellular functions. Biochem J, 2009. 423(2): p. 253-64.

[41] Anastasiou, D., et al., Inhibition of pyruvate kinase M2 by reactive oxygen species contributes to cellular antioxidant responses. Science, 2011. 334(6060): p. 1278-83.

[42] Kung, C., et al., Small molecule activation of PKM2 in cancer cells induces serine auxotrophy. Chem Biol, 2012. 19(9): p. 1187-98.

[43] DeNicola, G.M., et al., *NRF2 regulates serine biosynthesis in non-small cell lung cancer*. Nat Genet, 2015. 47(12): p. 1475-81.

[44] Maddocks, O.D., et al., Serine starvation induces stress and p53-dependent metabolic remodelling in cancer cells. Nature, 2013. 493(7433): p. 542-6.

[45] Cheung, E.C., et al., Opposing effects of TIGAR- and RAC1-derived ROS on Wnt-driven proliferation in the mouse intestine. Genes Dev, 2016. 30(1): p. 52-63.

[46] Bensaad, K., et al., TIGAR, a p53-inducible regulator of glycolysis and apoptosis. Cell, 2006. 126(1): p. 107-20.

[47] Bensaad, K. and K.H. Vousden, *p53: new roles in metabolism*. Trends Cell Biol, 2007. 17(6): p. 286-91.

[48] Li, H. and G. Jogl, Structural and biochemical studies of TIGAR (TP53-induced glycolysis and apoptosis regulator). J Biol Chem, 2009. 284(3): p. 1748-54.

[49] Wanka, C., J.P. Steinbach, and J. Rieger, Tp53-induced glycolysis and apoptosis regulator (TIGAR) protects glioma cells from starvation-induced cell death by up-regulating respiration and improving cellular redox homeostasis. J Biol Chem, 2012. 287(40): p. 33436-46.

[50] Cheung, E.C., R.L. Ludwig, and K.H. Vousden, *Mitochondrial localization of TIGAR under hypoxia stimulates HK2 and lowers ROS and cell death*. Proc Natl Acad Sci U S A, 2012. 109(50): p. 20491-6.

[51] Mailloux, R.J., X. Jin, and W.G. Willmore, Redox regulation of mitochondrial function with emphasis on cysteine oxidation reactions. Redox Biol, 2014. 2: p. 123-39.

[52] Lee, G., et al., Oxidative dimerization of PHD2 is responsible for its inactivation and contributes to metabolic reprogramming via HIF-1alpha activation. Sci Rep, 2016. 6: p. 18928.

[53] Semenza, G.L., HIF-1 mediates metabolic responses to intratumoral hypoxia and oncogenic mutations. J Clin Invest, 2013. 123(9): p. 3664-71.

[54] Zhao, T., et al., HIF-1-mediated metabolic reprogramming reduces ROS levels and facilitates the metastatic colonization of cancers in lungs. Sci Rep, 2014. 4: p. 3793.

[55] Heiss, E.H., et al., Glucose availability is a decisive factor for Nrf2-mediated gene expression. Redox Biol, 2013. 1: p. 359-65.

[56] Patra, K.C. and N. Hay, *The pentose phosphate pathway and cancer*. Trends Biochem Sci, 2014. 39(8): p. 347-54.

[57] Levy, H.R., R.R. Raineri, and B.H. Nevaldine, *On the structure and catalytic function of mammary glucose 6-phosphate dehydrogenase*. J Biol Chem, 1966. 241(10): p. 2181-7.

[58] Singh, A., et al., Transcription factor NRF2 regulates miR-1 and miR-206 to drive tumorigenesis. J Clin Invest, 2013. 123(7): p. 2921-34.

[59] Mitsuishi, Y., et al., Nrf2 redirects glucose and glutamine into anabolic pathways in metabolic reprogramming. Cancer Cell, 2012. 22(1): p. 66-79.

[60] Futerman, A.H. and H. Riezman, *The ins and outs of sphingolipid synthesis*. Trends Cell Biol, 2005. 15(6): p. 312-8.

[61] Kuge, O. and M. Nishijima, Biosynthetic regulation and intracellular transport of phosphatidylserine in mammalian cells. J Biochem, 2003. 133(4): p. 397-403.

[62] Loenen, W.A., S-adenosylmethionine: jack of all trades and master of everything? Biochem Soc Trans, 2006. 34(Pt 2): p. 330-3.

[63] Teperino, R., K. Schoonjans, and J. Auwerx, Histone methyl transferases and demethylases; can they link metabolism and transcription? Cell Metab, 2010. 12(4): p. 321-7.

[64] de Koning, T.J., et al., *L-serine in disease and development*. Biochem J, 2003. 371(Pt 3): p. 653-61.

[65] Pompella, A., et al., *The changing faces of glutathione, a cellular protagonist*. Biochem Pharmacol, 2003. 66(8): p. 1499-503.

[66] Barker, G.A. and J.C. Ellory, *The identification of neutral amino acid transport systems*. Exp Physiol, 1990. 75(1): p. 3-26.

[67] Palacin, M., et al., Molecular biology of mammalian plasma membrane amino acid transporters. Physiol Rev, 1998. 78(4): p. 969-1054.

[68] Mullarky, E., et al., *PHGDH amplification and altered glucose metabolism in human melanoma*. Pigment Cell Melanoma Res, 2011. 24(6): p. 1112-5.

[69] Yueh, M.F. and R.H. Tukey, Nrf2-Keap1 signaling pathway regulates human UGT1A1 expression in vitro and in transgenic UGT1 mice. J Biol Chem, 2007. 282(12): p. 8749-58.

[70] Wu, K.C., J.Y. Cui, and C.D. Klaassen, Effect of graded Nrf2 activation on phase-I and -II drug metabolizing enzymes and transporters in mouse liver. PLoS One, 2012. 7(7): p. e39006.

[71] Thimmulappa, R.K., et al., Identification of Nrf2-regulated genes induced by the chemopreventive agent sulforaphane by oligonucleotide microarray. Cancer Res, 2002. 62(18): p. 5196-203.

[72] Hirotsu, Y., et al., Nrf2-MafG heterodimers contribute globally to antioxidant and metabolic networks. Nucleic Acids Res, 2012. 40(20): p. 10228-39.

[73] Averous, J., et al., Induction of CHOP expression by amino acid limitation requires both ATF4 expression and ATF2 phosphorylation. J Biol Chem, 2004. 279(7): p. 5288-97.

[74] Agyeman, A.S., et al., Transcriptomic and proteomic profiling of KEAP1 disrupted and sulforaphane-treated human breast epithelial cells reveals common expression profiles. Breast Cancer Res Treat, 2012. 132(1): p. 175-87.

[75] Slocum, S.L., et al., Keap1/Nrf2 pathway activation leads to a repressed hepatic gluconeogenic and lipogenic program in mice on a high-fat diet. Arch Biochem Biophys, 2016. 591: p. 57-65.

[76] Meakin, P.J., et al., Susceptibility of Nrf2-null mice to steatohepatitis and cirrhosis upon consumption of a high-fat diet is associated with oxidative stress, perturbation of the unfolded protein response, and disturbance in the expression of metabolic enzymes but not with insulin resistance. Mol Cell Biol, 2014. 34(17): p. 3305-20.

[77] Pang, S., et al., SKN-1 and Nrf2 couples proline catabolism with lipid metabolism during nutrient deprivation. Nat Commun, 2014. 5: p. 5048.

[78] Pike, L.S., et al., Inhibition of fatty acid oxidation by etomoxir impairs NADPH production and increases reactive oxygen species resulting in ATP depletion and cell death in human glioblastoma cells. Biochim Biophys Acta, 2011. 1807(6): p. 726-34.

[79] Marro, S., et al., Heme controls ferroportin1 (FPN1) transcription involving Bach1, Nrf2 and a MARE/ARE sequence motif at position −7007 of the FPN1 promoter. Haematologica, 2010. 95(8): p. 1261-8.

[80] Baranano, D.E., et al., *Biliverdin reductase: a major physiologic cytoprotectant*. Proc Natl Acad Sci U S A, 2002. 99(25): p. 16093-8.

[81] Takeda, T.A., et al., *Continuous de novo biosynthesis of haem and its rapid turnover to bilirubin are necessary for cytoprotection against cell damage*. Sci Rep, 2015. 5: p. 10488.

[82] Jain, A., et al., *p62/SQSTM1 is a target gene for transcription factor NRF2 and creates a positive feedback loop by inducing antioxidant response element-driven gene transcription*. J Biol Chem, 2010. 285(29): p. 22576-91.

[83] Taguchi, K., et al., *Keap1 degradation by autophagy for the maintenance of redox homeostasis*. Proc Natl Acad Sci U S A, 2012. 109(34): p. 13561-6.

[84] Komatsu, M., et al., *The selective autophagy substrate p62 activates the stress responsive transcription factor Nrf2 through inactivation of Keap1*. Nat Cell Biol, 2010. 12(3): p. 213-23.

[85] Ichimura, Y., et al., *Phosphorylation of p62 activates the Keap1-Nrf2 pathway during selective autophagy*. Mol Cell, 2013. 51(5): p. 618-31.

[86] Bae, S.H., et al., *Sestrins activate Nrf2 by promoting p62-dependent autophagic degradation of Keap1 and prevent oxidative liver damage*. Cell Metab, 2013. 17(1): p. 73-84.

[87] Wolfson, R.L., et al., *Sestrin2 is a leucine sensor for the mTORC1 pathway*. Science, 2016. 351(6268): p. 43-8.

[88] Arlt, A., et al., *Increased proteasome subunit protein expression and proteasome activity in colon cancer relate to an enhanced activation of nuclear factor E2-related factor 2 (Nrf2)*. Oncogene, 2009. 28(45): p. 3983-96.

[89] Kwak, M.K. and T.W. Kensler, *Induction of 26S proteasome subunit PSMB5 by the bifunctional inducer 3-methylcholanthrene through the Nrf2-ARE, but not the AhR/Arnt-XRE, pathway*. Biochem Biophys Res Commun, 2006. 345(4): p. 1350-7.

[90] Kwak, M.K., et al., *Antioxidants enhance mammalian proteasome expression through the Keap1-Nrf2 signaling pathway*. Mol Cell Biol, 2003. 23(23): p. 8786-94.

[91] Pickering, A.M., et al., *Nrf2-dependent induction of proteasome and Pa28alphabeta regulator are required for adaptation to oxidative stress*. J Biol Chem, 2012. 287(13): p. 10021-31.

[92] Kapeta, S., N. Chondrogianni, and E.S. Gonos, *Nuclear erythroid factor 2-mediated proteasome activation delays senescence in human fibroblasts*. J Biol Chem, 2010. 285(11): p. 8171-84.

[93] Telorack, M., et al., *A glutathione-Nrf2-thioredoxin cross-talk ensures keratinocyte survival and efficient wound repair*. PLoS Genet, 2016. 12(1): p. e1005800.

[94] Duru, N., et al., *NRF2/miR-140 signaling confers radioprotection to human lung fibroblasts*. Cancer Lett, 2015. 369(1): p. 184-91.

[95] Shibata, T., et al., *Cancer related mutations in NRF2 impair its recognition by Keap1-Cul3 E3 ligase and promote malignancy*. Proc Natl Acad Sci U S A, 2008. 105(36): p. 13568-73.

[96] Ohta, T., et al., *Loss of Keap1 function activates Nrf2 and provides advantages for lung cancer cell growth*. Cancer Res, 2008. 68(5): p. 1303-9.

[97] Cancer Genome Atlas Research, N., *Comprehensive genomic characterization of squamous cell lung cancers*. Nature, 2012. 489(7417): p. 519-25.

[98] Singh, A., et al., *Gain of Nrf2 function in non-small-cell lung cancer cells confers radioresistance*. Antioxid Redox Signal, 2010. 13(11): p. 1627-37.

[99] Locasale, J.W., et al., *Phosphoglycerate dehydrogenase diverts glycolytic flux and contributes to oncogenesis*. Nat Genet, 2011. 43(9): p. 869-74.

[100] Ahmad, F., et al., *Nrf2-driven TERT regulates pentose phosphate pathway in glioblastoma*. Cell Death Dis, 2016. 7: p. e2213.

[101] Xu, I.M., et al., *Transketolase counteracts oxidative stress to drive cancer development*. Proc Natl Acad Sci U S A, 2016. 113(6): p. E725-34.

[102] Wen, D., et al., *Malic enzyme 1 induces epithelial-mesenchymal transition and indicates poor prognosis in hepatocellular carcinoma*. Tumour Biol, 2015. 36(8): p. 6211-21.

[103] Son, J., et al., *Glutamine supports pancreatic cancer growth through a KRAS-regulated metabolic pathway*. Nature, 2013. 496(7443): p. 101-5.

[104] Jiang, L., et al., *Reductive carboxylation supports redox homeostasis during anchorage-independent growth*. Nature, 2016. 532(7598): p. 255-8.

[105] Saito, T., et al., *p62/Sqstm1 promotes malignancy of HCV-positive hepatocellular carcinoma through Nrf2-dependent metabolic reprogramming*. Nat Commun, 2016. 7: p. 12030.

[106] Gerschman, R., et al., *Oxygen poisoning and x-irradiation: a mechanism in common*. Science, 1954. 119(3097): p. 623-6.

[107] Vina, J., et al., *The free radical theory of aging revisited: the cell signaling disruption theory of aging*. Antioxid Redox Signal, 2013. 19(8): p. 779-87.

[108] Aydin, A.F., et al., *The effect of carnosine treatment on prooxidant-antioxidant balance in liver, heart and brain tissues of male aged rats*. Biogerontology, 2010. 11(1): p. 103-9.

[109] Legan, S.K., et al., *Overexpression of glucose-6-phosphate dehydrogenase extends the life span of Drosophila melanogaster*. J Biol Chem, 2008. 283(47): p. 32492-9.

[110] Nobrega-Pereira, S., et al., *G6PD protects from oxidative damage and improves healthspan in mice*. Nat Commun, 2016. 7: p. 10894.

[111] Hirota, A., et al., *Acceleration of UVB-induced photoageing in nrf2 gene-deficient mice*. Exp Dermatol, 2011. 20(8): p. 664-8.

[112] Hoshino, T., et al., *Protective role of Nrf2 in age-related hearing loss and gentamicin ototoxicity*. Biochem Biophys Res Commun, 2011. 415(1): p. 94-8.

[113] Zhang, H., K.J. Davies, and H.J. Forman, *Oxidative stress response and Nrf2 signaling in aging*. Free Radic Biol Med, 2015. 88(Pt B): p. 314-36.

[114] Shenvi, S.V., E. Smith, and T.M. Hagen, *Identification of age-specific Nrf2 binding to a novel antioxidant response element locus in the Gclc promoter: a compensatory means for the loss of glutathione synthetic capacity in the aging rat liver?* Aging Cell, 2012. 11(2): p. 297-304.

[115] Kubben, N., et al., *Repression of the antioxidant NRF2 pathway in premature aging*. Cell, 2016. 165(6): p. 1361-74.

[116] Scaffidi, P. and T. Misteli, *Lamin A-dependent misregulation of adult stem cells associated with accelerated ageing*. Nat Cell Biol, 2008. 10(4): p. 452-9.

[117] Bigarella, C.L., R. Liang, and S. Ghaffari, *Stem cells and the impact of ROS signaling*. Development, 2014. 141(22): p. 4206-18.

[118] Kobayashi, C.I. and T. Suda, *Regulation of reactive oxygen species in stem cells and cancer stem cells*. J Cell Physiol, 2012. 227(2): p. 421-30.

[119] Cervantes, R.B., et al., *Embryonic stem cells and somatic cells differ in mutation frequency and type*. Proc Natl Acad Sci U S A, 2002. 99(6): p. 3586-90.

[120] Saretzki, G., et al., *Stress defense in murine embryonic stem cells is superior to that of various differentiated murine cells*. Stem Cells, 2004. 22(6): p. 962-71.

[121] Folmes, C.D., et al., *Somatic oxidative bioenergetics transitions into pluripotency-dependent glycolysis to facilitate nuclear reprogramming*. Cell Metab, 2011. 14(2): p. 264-71.

[122] Panopoulos, A.D., et al., *The metabolome of induced pluripotent stem cells reveals metabolic changes occurring in somatic cell reprogramming*. Cell Res, 2012. 22(1): p. 168-77.

[123] Hochmuth, C.E., et al., *Redox regulation by Keap1 and Nrf2 controls intestinal stem cell proliferation in Drosophila*. Cell Stem Cell, 2011. 8(2): p. 188-99.

[124] Merchant, A.A., et al., *The redox-sensitive transcription factor Nrf2 regulates murine hematopoietic stem cell survival independently of ROS levels*. Blood, 2011. 118(25): p. 6572-9.

[125] Pan, H., et al., *SIRT6 safeguards human mesenchymal stem cells from oxidative stress by coactivating NRF2*. Cell Res, 2016. 26(2): p. 190-205.

[126] Corenblum, M.J., et al., *Reduced Nrf2 expression mediates the decline in neural stem cell function during a critical middle-age period*. Aging Cell, 2016. 15(4): p. 725-36.

[127] Hawkins, K.E., et al., *NRF2 orchestrates the metabolic shift during induced pluripotent stem cell reprogramming*. Cell Rep, 2016. 14(8): p. 1883-91.

[128] Dannenmann, B., et al., *High glutathione and glutathione peroxidase-2 levels mediate cell-type-specific DNA damage protection in human induced pluripotent stem cells*. Stem Cell Rep, 2015. 4(5): p. 886-98.

[129] Zhu, J., et al., *Nrf2 is required to maintain the self-renewal of glioma stem cells*. BMC Cancer, 2013. 13: p. 380.

[130] Pan, J., et al., *26S proteasome activity is down-regulated in lung cancer stem-like cells propagated in vitro*. PLoS One, 2010. 5(10): p. e13298.

Chapter 3

Multiple Modes of Nrf2 Regulation and Transcriptional Response

Sherin T. Mathew and Ola Hammarsten

Additional information is available at the end of the chapter

http://dx.doi.org/10.5772/65295

Abstract

Cells have defense systems to deal with chemical insults from the environment. Some examples are chemical scavengers like glutathione and enzymes such as superoxide dismutase that inactivate radicals and other reactive chemicals in the cytoplasm. It is perhaps surprising that these protective systems are not maximally expressed in an unstressed cell. Rather, the ability to inactivate toxic chemicals is tightly regulated and only induced when needed. As a consequence, unstressed cells are usually very sensitive to radicals, but become more resistant as the cellular defense system has been appropriately upregulated after a few hours. The transcription factor Nrf2 is known to be a master regulator of many cytoprotective enzymes and proteins. Chemical inducers of Nrf2 inactivate its repressor, Keap1, when they react with critical cysteine residues in Keap1. The release of Nrf2 from Keap1 results in enhanced expression of genes involved in detoxification. This generates a feedback loop where Nrf2 induces protective enzymes capable of inactivating the chemical that reacted with Keap1. An unproven, but likely, scenario is that Nrf2 transcriptional response can vary depending on the nature of the chemical insult. The aim of this chapter is to examine the mechanisms by which the cell can sense different reactive chemicals and modulate protective responses. It is likely that this knowledge is of vital importance in the development of clinical Nrf2 activators in preventive medicine.

Keywords: Keap1, Nrf2, cysteine residues, transcriptional response, Nrf2 activator

1. Introduction

Cells in our bodies are constantly challenged by reactive substances such as radicals, in part due to the high oxygen content in the surrounding. To prevent unwanted chemical reactions, cells have evolved sophisticated defense mechanisms. These systems allow cellular adaptation to toxins in our environment by regulating the expression of an elaborate network of cytoprotective proteins and chemical scavengers.

To a large extent, this cytoprotective system is regulated by the transcription factor nuclear E2-factor-related factor 2 (Nrf2) and its primary suppressor kelch-like ECH-associated protein 1 (Keap1). Under basal conditions, Keap1 decreases the Nrf2 protein levels through ubiquitination and proteasomal degradation [1, 2]. Upon activation, Nrf2 escapes Keap1 repression and Nrf2 protein levels get stabilized and translocate to the nucleus. In the nucleus, Nrf2-Maf heterodimer binds to the antioxidant response element (ARE) in the regulatory region of target genes and promotes transcription [3]. A wide array of genes including antioxidant proteins and detoxifying enzymes, transporter and metabolic enzymes, enzymes for glutathione biosynthesis, proteases, and chaperone are transcriptionally activated by Nrf2 [4]. It is likely that the cellular response to a chemical challenge will be different if the toxic substance is an oxidative or a reductive toxin.

An example of this diversification in response can be found in the bacteria *Escherichia coli* where the redox system is mainly regulated by transcription factors OxyR and SoxRS. The cysteine residues of OxyR sense elevated levels of hydrogen peroxide [5], whereas SoxRS iron-sulfur (2Fe-2S) clusters act as sensors of superoxide. When activated in this way, OxyR and SoxRS become active transcription factors that induce distinct but partially overlapping set of cytoprotective genes adapting the cell to deal with hydrogen peroxide or superoxide.

In a similar way, it is thought that our cells diversify the redox response depending on the nature of the chemical insult. As mentioned above, electrophiles and oxidants react with cysteine residues in Keap1 which blocks Nrf2 proteasomal degradation and thus mediates transcriptional activation of many genes [6]. However, little is known about how the Nrf2 system can differentiate the transcriptional response. Here, we review the most recent literature on how Nrf2 cross talks with multiple signaling pathways and evokes different signaling response including inflammation, metabolism, apoptosis, proliferation, and differentiation. This knowledge is likely of great importance when Nrf2-activating drugs are developed to boost our radical defense systems.

2. Redox regulation in bacteria

The bacterial redox system in *E. coli* is coordinately regulated by two transcription factors—SoxRS and OxyR. SoxRS is activated in response to stress induced by superoxide anion and OxyR responds to stress caused by hydrogen peroxide.

2.1. OxyR-hydrogen peroxide sensor

The transcription factor OxyR belongs to LysR family of transcriptional activators. In response to hydrogen peroxide stress, two cysteine residues, Cys-199 and Cys-208 oxidize leading to the formation of intramolecular disulfide bond [7]. Oxidized OxyR promotes transcription of genes including *katG* (a hydrogen peroxidase I), *ahpCF* (an alkylhydroperoxide reductase), *oxyS* (a regulatory RNA involved in DNA repair), *gorA* (a glutathione reductase), and glutaredoxin 1 (*grxA*). Enzymatic reduction of the disulfide bond switches off the OxyR function and the OxyR transcription factor therefore functions as an "on/off switch" with disulfide bond formation in response to oxidative stress [8–10]. Studies suggest that Cys-199 thiol activates OxyR through several redox-related modifications including S-OH, S-nitrosylation, and S-glutathione and the resulting OxyR differs in structure, properties, and genes activated [5].

2.2. SoxRS-superoxide sensor

The superoxide response system in *E. coli* regulates transcription of targets involved in detoxification (superoxide dismutase), DNA repair (endonuclease IV), and glucose-6-phosphate dehydrogenase. The iron-sulfur (2Fe-2S) clusters of the SoxR act as sensors of superoxide and undergo one-electron oxidation/reduction and induce the transcriptional activity of SoxR [11]. SoxR protein binds DNA to activate the expression of SoxS which in turn activates cell protection genes in response to superoxide and nitric oxide [12]. Site-specific mutation studies have shown that four conserved cysteine residues at the C-terminal domain of the SoxR polypeptide act as the ligands for the [2Fe-2S] clusters which has crucial role in transcriptional activity of SoxR [13].

2.3. OxyR-SoxR interaction

Both OxyR and SoxR proteins exist in oxidized and reduced forms but only the oxidized form of these proteins induces the expression of antioxidant defense system. Although the SoxR and SoxS proteins are mainly involved in response to superoxide, several studies have reported that SoxRS regulon may be activated by hydrogen peroxide indicating the overlapping between the specific response systems. SoxR protein senses the increased levels of hydrogen peroxide and activates the SoxRS system [14–16].

3. Keap1-mediated Nrf2 regulation

The critical importance of Keap1 as a negative regulator of Nrf2 is supported by the observation that the deletion of *Keap1* gene in mice causes constitutive activation of Nrf2. Keap1$^{-/-}$ knockout mice died shortly after birth due to hyperkeratosis in the upper digestive tract but the phenotype conditions were reversed when both Nrf2 and Keap1 were disrupted [17, 18]. The cysteine-rich protein Keap1 regulates active degradation of Nrf2 under basal conditions by functioning as an adaptor to cullin3 (Cul3)-ringbox1 (Rbx1) containing E3 ubiquitin ligase complex (**Figure 1**) [19]. The Neh2 domain of Nrf2 binds to Kelch domain of Keap1 through the "hinge-and-latch" mechanism [20]. Under basal conditions, Nrf2 ETGE motif acts as a hinge and forms an "open" conformation by binding to Kelch subunit of Keap1 and the DLG motif which acts as the latch binds to Keap1 subunit to form the "closed" conformation and targets Nrf2 for proteasomal degradation. Cysteine residues of Keap1 sense reactive oxygen species (ROS) or electrophiles in the cellular environment causing conformation changes in

Keap1. The modified Keap1 can disrupt its interaction with the low-affinity DLG motif, whereas the high-affinity ETGE motif remains associated with Keap1. As the DLG motif fails to bind to Keap1, it affects the orientation of lysine residues within the Neh2 domain of Nrf2 preventing its ubiquitination and degradation [21]. After redox homeostasis is restored, Keap1 moves into the nucleus and controls nuclear export of Nrf2 for subsequent proteasomal degradation in the cytoplasm (**Figure 1**) [22].

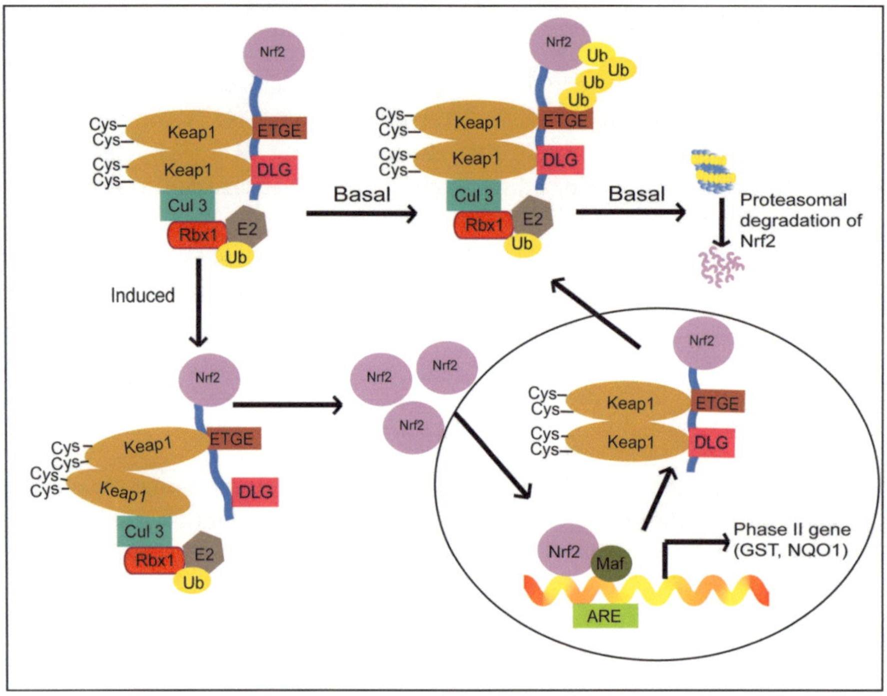

Figure 1. Keap1-mediated Nrf2 regulatory pathway. Under basal conditions, Nrf2 is bound to Keap1 and undergoes rapid degradation. Upon induction, cysteine residues in Keap1 are modified, the E3 ubiquitin ligase activity is suppressed, and Nrf2 levels increase. Activated Nrf2 enters the nucleus and dimerizes with Maf to promote transcription of ARE-dependent genes. Finally, Nrf2 is transported out of the nucleus by Keap1 for subsequent proteasomal degradation.

3.1. Distinct Keap1 cysteine modifications

Various synthetic and plant-derived phytochemicals including isothiocyanates, Michael acceptors, and coumarins are shown to activate Nrf2 system [23]. Many of these substances protect human cells and animals from a diverse array of toxins and radiation [24]. These structurally diverse Nrf2-inducing agents share a common property of reacting with sulfhydryl groups and cysteine residues in Keap1 [6, 25]. ROS or electrophilic reaction with specific cysteine residues causes conformational changes in Keap1 and prevents proteasomal degradation of Nrf2 [26]. Site-directed mutagenic studies have identified critical cysteine residues as important factors involved in Nrf2 regulation. Mutation in Cys273 or Cys288, located at intervening region of Keap1, blocked the Keap1-dependent ubiquitination and degradation of Nrf2 under basal conditions. Mutation in Cys151 at the BTB domain of Keap1 blocked Nrf2 release from Keap1 in response to sulforaphane and maintained Keap1 repression of Nrf2. Cys273 or Cys288 is required for Keap1 repression of Nrf2 under basal conditions and Cys151 is important for Nrf2 activation in response to electrophilic stress [27–29].

A "cysteine code" has been proposed for the Keap1-dependent Nrf2 regulation as cysteine modifications play crucial role in mediating Nrf2 activation [30]. Studies suggest that Nrf2-inducing agents such as diethylmaleate (DEM), dimethylfumarate (DMF), and sulforaphane prefer Cys151 residue for Nrf2 induction, whereas 2-cyano-3, 12 dioxooleana-1, 9d iene-28-imidazolide (CDDO-Im) and heavy metals such as cadmium chloride (CdCl2) and arsenic activate Nrf2 in a Cys151-independent manner. Another well-known Nrf2 inducer, tert-butyl hydroquinone (tBHQ), gets oxidized to the electrophilic metabolite tert-butyl benzoquinone and modifies the Cys151 cysteine residues [31, 32]. The differential reactivity of cysteines in Keap1, "the cysteine code," does not, however, explain how this translates into differential toxin-dependent activation of genes by Nrf2.

4. Nrf2 network

Recent advances have revealed that Nrf2 cross talks with different signaling pathways and influences the transcriptional response. Beyond cellular response against oxidative stress, Nrf2 is reported to be involved in inflammation, metabolism, apoptosis, proliferation, and differentiation.

4.1. Cross talk between Nrf2 and NF-κB pathway

Nuclear factor-κB (NF-κB) was first identified in David Baltimore's laboratory around 30 years ago as a transcription factor that activated the kB immunoglobulin promoter in B-cells [33]. Since then, NF-κB has been implemented in a diverse array of conditions mostly linked to acute and chronic inflammation.

NF-κB is normally sequestered in the cytoplasm by its negative regulator IkB-alpha;. Upon T-cell or B-cell receptor activation in response to infection or tumor necrosis factor (TNF)α receptor stimulation IkBα is phosphorylated by the IkB kinase (IKK) complex. Phosphorylation of *Ik*Bα is followed by ubiquitination and proteasomal degradation of IkBα and releases NF-κB. NF-κB translocates to the nucleus and activates target genes having the kB elements in their promoters [34]. It has been proposed that after simultaneous activation, NF-κB antagonizes Nrf2-mediated gene transcription. Conversely, some Nrf2 inducers suppress NF-κB signaling. Therefore, it seems that the inflammatory induction of NF-κB can be suppressed by Nrf2 activation and vice versa. It is possible that the inflammatory response that generates radicals to defeat bacteria must downregulate the radical scavenger function of the Nrf2 to function optimally [35] (**Figure 2**). Similarly, the radicals produced by the inflammatory response could activate the Nrf2 system in neighboring normal cells.

Therefore, the anti-inflammatory effects of Nrf2 can be due, in part, to its ability to act as a feedback regulator of NF-κB and the absence of Nrf2 can create a situation where NF-κB lacks a controller to turn off the inflammatory signal, resulting in chronic inflammatory conditions such as observed in arteries damaged by ionizing radiation [36]. For example, Nrf2 knockout mice are more susceptible to lipopolysaccharide (LPS)-induced neuroinflammation. Activation of the Nrf2 pathway in normal cells with sulforaphane decreased the production of inflammatory markers [37]. The Nrf2 target gene hemeoxygenase-1 (HO-1) inhibits NF-κB-mediated transcription of cellular adhesion molecules and could thereby block accumulation of inflammatory cells [38]. In addition, Keap1 downregulates NF-κB by promoting proteasomal degradation of its activator IKKβ [39].

Similarly, there are several examples of how NF-κB downregulates the Nrf2 response. Keap1 interacts with NF-κB, and thereby represses the Nrf2 transcriptional activity [40]. NF-κB blocks Nrf2 transcriptional activation of target genes [41]. This may result in increased oxidative stress which in turn further activates NF-κB [42]. In addition, NF-κB can also promote HDAC3 association with MafK and thus compete with Nrf2 heterodimer formation and transcription of Nrf2-dependent genes [43].

There are also examples where NF-κB promoting activation of Nrf2 and Nrf2-regulated genes. Functional NF-κB-binding sites have been found in the promoter of the NRF2 gene resulting in overexpression of Nrf2 in acute myeloid leukemia cells [44]. In another study, the activation of small GTPase protein RAC1 (Ras-related C3 botulinum toxin substrate 1) induced the expression of Nrf2 target gene HO-1 and caused the inhibition of NF-κB function [45].

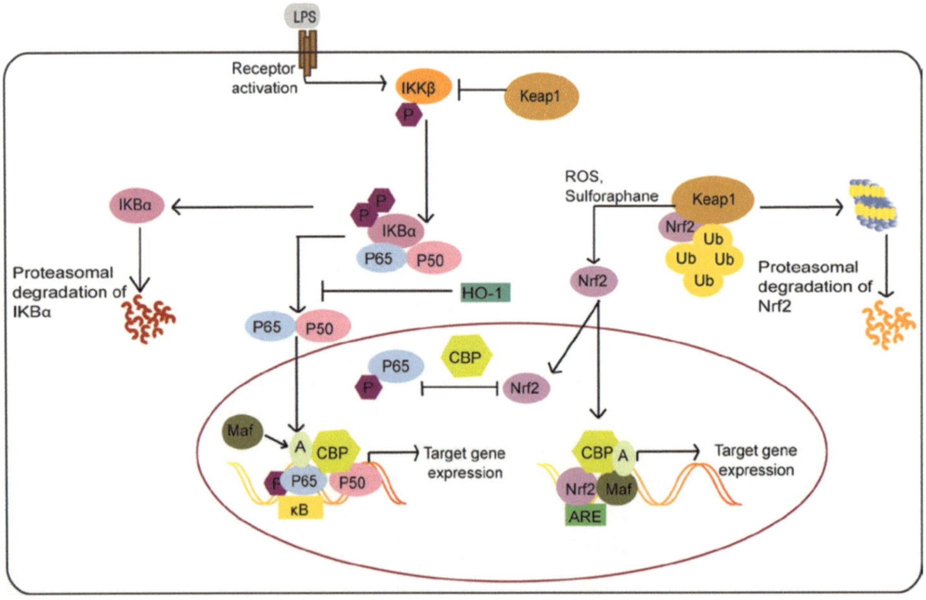

Figure 2. Different factors involved in the functional network between Nrf2 and NF-κB network.

Several other proteins are known to interact with both pathways during the signaling process. One example is p62 protein which accumulates due to autophagy deficiency, and activates Nrf2 through direct interaction with Keap1. P62 sequesters Keap1 into aggregates and inhibits Keap1-mediated ubiquitylation and degradation of Nrf2. This resulted in increased Nrf2 stabilization and activation of target genes [46, 47]. Similarly, p62 protein oligomerizes and promotes nerve growth factor (NGF)-mediated NF-κB signaling [48].

Another protein that interacts with both Nrf2 and NF-κB pathway is glycogen synthase kinase-3β (GSK-3β), a Ser/Thr kinase involved in glycogen metabolism and apoptosis. GSK-3β phosphorylates the Neh6 domain of Nrf2 and targets subsequent proteasomal degradation by β-TrCP (β-transducing repeat-containing protein)-Skp1(S-phase kinase-associated protein1)-Cul1-Rbx1 E3 ubiquitin ligase complex [49]. In NF-κB system, GSK-3*β* phosphorylates p65 subunit and increases its DNA-binding affinity and subsequent transcriptional response [50]. Moreover, *β*-TrCP mediates proteasomal degradation of the inhibitory protein, I*k*B*α*, and allows NF-κB release. Thus, *β*-TrCP functions as positive regulator of NF-κB activity and negative regulator of Nrf2-ARE activity (**Figure 2**).

4.2. Nrf2 and other signaling pathways

Phosphorylation of Nrf2 by several protein kinases can lead to stabilization and activation of Nrf2. For example, several studies have demonstrated the involvement of P13 kinase/AKT pathway in regulating Nrf2 nuclear translocation and ARE-dependent gene expression [51, 52]. P13K phosphorylation of PKB/Akt suppresses proteasomal degradation of Nrf2 by GSK-3β [53]. In another study, the biotinylated derivative of the triterpenoid CDDO (2-cyano-3,12-dioxooleana-1,9-dien-28-oic acid) activated PI3K-PKB/Akt signaling through modification of Cys-124 in the active site of PTEN, causing inhibition of the lipid phosphatase function of PTEN [54]. Similarly, increased Nrf2 activity was observed in PTEN-mutant cells with increased activation of the PI3K-Akt signaling which in turn suppressed the GSK-3β-mediated Nrf2 repression [55]. But in Keap1-deficient cells, the deletion of PTEN increased Nrf2 accumulation to a greater extent and this supports the regulatory role of PTEN-GSK-3-β-TrCP signaling [56]. Under conditions of autophagy dysregulation, Nrf2 is activated by p62. In a recent study, Nrf2 inducer sulforaphane activated p62 through SPBP (stromelysin-1 platelet-derived growth factor-responsive element-binding protein), which acts as a coactivator of Nrf2 [57]. Another Nrf2 activator, arsenic, activates Nrf2 through p62-mediated Keap1 sequestration and this persistent activation of Nrf2 can be the reason for arsenic-induced toxicity [58]. The mitogen-activated protein kinase (MAPK)-signaling pathway is activated by Nrf2 inducers like tBHQ and sulforaphane [59, 60]. P38, a member of MAPK family, has been shown to influence Nrf2 activation both in positive and negative manner and this suggests the complex nature of Nrf2 regulation by MAPK [60]. During endoplasmic reticulum stress, protein kinase PERK activates Nrf2 and provides cell survival benefits. PERK-induced phosphorylation of Nrf2 allows its release from Keap1 and translocation into the nucleus for subsequent gene transcription [61].

Another signaling pathway involving protein kinase C has been shown to increase Nrf2-target gene expression while its inhibition caused significant decrease in tBHQ-induced Nrf2 nuclear translocation [62]. PKC phosphorylates serine 40 of Nrf2 resulting in the Nrf2 release from Keap1 for subsequent gene expressions and mutation of specific residue reduced the gene expression level by 50% indicating that PKC functions along with Keap1 [63, 64].

5. Differential Nrf2 response

Based on the evidences, different inducers target specific or combination of different cysteine residues to activate Nrf2, suggesting the function of "cysteine codes" which converts the preferential target cysteine modifications into distinct biological effects. Understanding the cysteine code for each Nrf2-activating compound will help to increase the biological effects of different inducers. However, Nrf2 activation could not be solely responsible for the diverse biological effects caused by Nrf2 inducers. Most of the Nrf2-inducing agents have the inherent ability to react with cysteine residues and there is a possibility of interaction with other cellular proteins and thereby generating distinct cellular response. For example, a proteomics study of sulforaphane-derived sulfoxythiocarbamate analogs has identified different protein targets other than Keap1 [65]. Many Nrf2 inducers are well known for the prevention and treatment of several human disorders and some of them have been clinically investigated. For example, the methyl ester derivative of CDDO triterpenoid is a potent Nrf2 inducer and as low nanomolar concentrations of CDDO-Me (bardoxolone methyl) stimulate Nrf2-dependent gene expressions. The phase II clinical trials for the treatment of chronic kidney disease (BEAM study) in patient with Type 2 diabetes indicated that CDDO-Me could improve kidney function. However, the phase III trial (BEACON study) was terminated due to serious side effects and mortality observed in treated group [66]. However, the exact mechanism behind the adverse effects are not clear, the long-term drug exposure as well as administration of fixed dose of drug not adjusted for kidney function might have influenced the response. Nrf2 is overexpressed in many types of cancers and several oncogenes are reported to evoke Nrf2 expression in cancer cells [67]. As Nrf2 plays a key role in cytoprotection, cancer cells benefit the protective effect of Nrf2 to create a favorable microenvironment for tumor growth and drug resistance. Studies found that two different antioxidants, N-acetylcysteine (NAC) or vitamin E supplementation in mice with lung tumors substantially increased the number, size, and stage of the tumors. NAC and vitamin E reduced the ROS levels, DNA damage, and expression of *p53* tumor-suppressor gene [68]. Similarly, NAC and vitamin E supplementation increased metastasis in mice with malignant melanoma [69]. Consistent results were observed in another independent study where NAC administration increased metastasis of melanoma in mice [70]. It is notable that ROS plays both negative and positive roles in cellular signaling. In normal conditions, low levels of ROS may function as messengers in cell signaling, while excess ROS levels have adverse effect on cellular macromolecules and lead to cell death. Antioxidant agents protect cells against ROS by increasing the antioxidant potential through Nrf2 activation. However, suppressing normal physiological ROS may affect the cell communication and signal transduction. For example, exercise generates ROS but promotes health benefits, especially in increasing insulin sensitivity. Transient production of ROS during exercise induced signaling systems that activated molecular targets for insulin sensitivity;

however, antioxidant supplementation blocked the molecular signaling for cellular defense and insulin sensitivity mediated by exercise-induced ROS formation and thereby abrogates the health-promoting effects of exercise [71].

Several complex transcriptional and posttranslational networks are involved in mediating Nrf2 activation and thereby enabling diverse functional response. Moreover, networking with other signaling pathways expands the function of Nrf2 as a potent regulator of differential biological processes such as cell proliferation, apoptosis, angiogenesis, and metastasis. The cross-talks between different transcription factors may influence the outcome of therapeutic interventions. Understanding the molecular mechanisms involved in regulating Nrf2 can therefore provide insights that may benefit novel therapeutic manipulation of this pathway.

Author details

Sherin T. Mathew and Ola Hammarsten*

*Address all correspondence to: ola.hammarsten@gu.se

Department of Clinical Chemistry, Transfusion Medicine, Institute of Biomedicine, Sahlgrenska Academy, University of Gothenburg, Gothenburg, Sweden

References

[1] Itoh K, Wakabayashi N, Katoh Y, Ishii T, Igarashi K, Engel JD, et al. Keap1 represses nuclear activation of antioxidant responsive elements by Nrf2 through binding to the amino-terminal Neh2 domain. Genes and Development. 1999;13(1):76–86.

[2] McMahon M, Itoh K, Yamamoto M, Hayes JD. Keap1-dependent proteasomal degradation of transcription factor Nrf2 contributes to the negative regulation of antioxidant response element-driven gene expression. The Journal of Biological Chemistry. 2003;278(24):21592–600.

[3] Itoh K, Chiba T, Takahashi S, Ishii T, Igarashi K, Katoh Y, et al. An Nrf2/small Maf heterodimer mediates the induction of phase II detoxifying enzyme genes through antioxidant response elements. Biochemical and Biophysical Research Communications. 1997;236(2):313–22.

[4] Kobayashi M, Yamamoto M. Molecular mechanisms activating the Nrf2-Keap1 pathway of antioxidant gene regulation. Antioxidants and Redox Signaling. 2005;7(3–4):385–94.

[5] Kim SO, Merchant K, Nudelman R, Beyer WF, Jr., Keng T, DeAngelo J, et al. OxyR: a molecular code for redox-related signaling. Cell. 2002;109(3):383–96.

[6] Dinkova-Kostova AT, Holtzclaw WD, Cole RN, Itoh K, Wakabayashi N, Katoh Y, et al. Direct evidence that sulfhydryl groups of Keap1 are the sensors regulating induction of phase 2 enzymes that protect against carcinogens and oxidants. Proceedings of the National Academy of Sciences of the United States of America. 2002;99(18):11908–13.

[7] Zheng M, Aslund F, Storz G. Activation of the OxyR transcription factor by reversible disulfide bond formation. Science. 1998;279(5357):1718–21.

[8] Lee C, Lee SM, Mukhopadhyay P, Kim SJ, Lee SC, Ahn WS, et al. Redox regulation of OxyR requires specific disulfide bond formation involving a rapid kinetic reaction path. Nature Structural and Molecular Biology. 2004;11(12):1179–85.

[9] Helmann JD. OxyR: a molecular code for redox sensing? Science's STKE. 2002;2002(157):pe46.

[10] Choi H, Kim S, Mukhopadhyay P, Cho S, Woo J, Storz G, et al. Structural basis of the redox switch in the OxyR transcription factor. Cell. 2001;105(1):103–13.

[11] Ding H, Hidalgo E, Demple B. The redox state of the [2Fe-2S] clusters in SoxR protein regulates its activity as a transcription factor. The Journal of Biological Chemistry. 1996;271(52):33173–5.

[12] Demple B, Amabile-Cuevas CF. Redox redux: the control of oxidative stress responses. Cell. 1991;67(5):837–9.

[13] Bradley TM, Hidalgo E, Leautaud V, Ding H, Demple B. Cysteine-to-alanine replacements in the Escherichia coli SoxR protein and the role of the [2Fe-2S] centers in transcriptional activation. Nucleic Acids Research. 1997;25(8):1469–75.

[14] Manchado M, Michan C, Pueyo C. Hydrogen peroxide activates the SoxRS regulon in vivo. Journal of Bacteriology. 2000;182(23):6842–4.

[15] Semchyshyn H, Bagnyukova T, Storey K, Lushchak V. Hydrogen peroxide increases the activities of soxRS regulon enzymes and the levels of oxidized proteins and lipids in *Escherichia coli*. Cell Biology International. 2005;29(11):898–902.

[16] Zheng M, Wang X, Templeton LJ, Smulski DR, LaRossa RA, Storz G. DNA microarray-mediated transcriptional profiling of the *Escherichia coli* response to hydrogen peroxide. Journal of Bacteriology. 2001;183(15):4562–70.

[17] Wakabayashi N, Itoh K, Wakabayashi J, Motohashi H, Noda S, Takahashi S, et al. Keap1-null mutation leads to postnatal lethality due to constitutive Nrf2 activation. Nature Genetics. 2003;35(3):238–45.

[18] Taguchi K, Maher JM, Suzuki T, Kawatani Y, Motohashi H, Yamamoto M. Genetic analysis of cytoprotective functions supported by graded expression of Keap1. Molecular and Cellular Biology. 2010;30(12):3016–26.

[19] Zhang DD, Lo SC, Cross JV, Templeton DJ, Hannink M. Keap1 is a redox-regulated substrate adaptor protein for a Cul3-dependent ubiquitin ligase complex. Molecular and Cellular Biology. 2004;24(24):10941–53.

[20] Tong KI, Kobayashi A, Katsuoka F, Yamamoto M. Two-site substrate recognition model for the Keap1-Nrf2 system: a hinge and latch mechanism. Biological Chemistry. 2006;387(10–11):1311–20.

[21] Tong KI, Katoh Y, Kusunoki H, Itoh K, Tanaka T, Yamamoto M. Keap1 recruits Neh2 through binding to ETGE and DLG motifs: characterization of the two-site molecular recognition model. Molecular and Cellular Biology. 2006;26(8):2887–900.

[22] Sun Z, Zhang S, Chan JY, Zhang DD. Keap1 controls postinduction repression of the Nrf2-mediated antioxidant response by escorting nuclear export of Nrf2. Molecular and Cellular Biology. 2007;27(18):6334–49.

[23] Kensler TW, Wakabayashi N. Nrf2: friend or foe for chemoprevention? Carcinogenesis. 2010;31(1):90–9.

[24] Kim SB, Pandita RK, Eskiocak U, Ly P, Kaisani A, Kumar R, et al. Targeting of Nrf2 induces DNA damage signaling and protects colonic epithelial cells from ionizing radiation. Proceedings of the National Academy of Sciences of the United States of America. 2012;109(43):E2949–55.

[25] Dinkova-Kostova AT, Talalay P. Direct and indirect antioxidant properties of inducers of cytoprotective proteins. Molecular Nutrition and Food Research. 2008;52(Suppl 1):S128–38.

[26] Li Y, Paonessa JD, Zhang Y. Mechanism of chemical activation of Nrf2. PLoS One. 2012;7(4):e35122.

[27] Zhang DD, Hannink M. Distinct cysteine residues in Keap1 are required for Keap1-dependent ubiquitination of Nrf2 and for stabilization of Nrf2 by chemopreventive agents and oxidative stress. Molecular and Cellular Biology. 2003;23(22):8137–51.

[28] Wakabayashi N, Dinkova-Kostova AT, Holtzclaw WD, Kang MI, Kobayashi A, Yamamoto M, et al. Protection against electrophile and oxidant stress by induction of the phase 2 response: fate of cysteines of the Keap1 sensor modified by inducers. Proceedings of the National Academy of Sciences of the United States of America. 2004;101(7): 2040–5.

[29] Yamamoto T, Suzuki T, Kobayashi A, Wakabayashi J, Maher J, Motohashi H, et al. Physiological significance of reactive cysteine residues of Keap1 in determining Nrf2 activity. Molecular and Cellular Biology. 2008;28(8):2758–70.

[30] He X, Ma Q. NRF2 cysteine residues are critical for oxidant/electrophile-sensing, Kelch-like ECH-associated protein-1-dependent ubiquitination-proteasomal degradation, and transcription activation. Molecular Pharmacology. 2009;76(6):1265–78.

[31] Takaya K, Suzuki T, Motohashi H, Onodera K, Satomi S, Kensler TW, et al. Validation of the multiple sensor mechanism of the Keap1-Nrf2 system. Free Radical Biology and Medicine. 2012;53(4):817–27.

[32] Wang XJ, Sun Z, Chen W, Li Y, Villeneuve NF, Zhang DD. Activation of Nrf2 by arsenite and monomethylarsonous acid is independent of Keap1-C151: enhanced Keap1-Cul3 interaction. Toxicology and Applied Pharmacology. 2008;230(3):383–9.

[33] Sen R, Baltimore D. Multiple nuclear factors interact with the immunoglobulin enhancer sequences. Cell. 1986;46(5):705–16.

[34] Winston JT, Strack P, Beer-Romero P, Chu CY, Elledge SJ, Harper JW. The SCF beta-TRCP-ubiquitin ligase complex associates specifically with phosphorylated destruction motifs in IkappaBalpha and beta-catenin and stimulates IkappaBalpha ubiquitination in vitro. Genes and Development. 1999;13(3):270–83.

[35] Buelna-Chontal M, Zazueta C. Redox activation of Nrf2 & NF-kappaB: a double end sword? Cell Signal. 2013;25(12):2548–57.

[36] Halle M, Gabrielsen A, Paulsson-Berne G, Gahm C, Agardh HE, Farnebo F, et al. Sustained inflammation due to nuclear factor-kappa B activation in irradiated human arteries. Journal of the American College of Cardiology. 2010;55(12):1227–36.

[37] Innamorato NG, Rojo AI, Garcia-Yague AJ, Yamamoto M, de Ceballos ML, Cuadrado A. The transcription factor Nrf2 is a therapeutic target against brain inflammation. Journal of Immunology. 2008;181(1):680–9.

[38] Soares MP, Seldon MP, Gregoire IP, Vassilevskaia T, Berberat PO, Yu J, et al. Heme oxygenase-1 modulates the expression of adhesion molecules associated with endothelial cell activation. Journal of Immunology. 2004;172(6):3553–63.

[39] Lee DF, Kuo HP, Liu M, Chou CK, Xia W, Du Y, et al. KEAP1 E3 ligase-mediated downregulation of NF-kappaB signaling by targeting IKKbeta. Molecular Cell. 2009;36(1):131–40.

[40] Yu M, Li H, Liu Q, Liu F, Tang L, Li C, et al. Nuclear factor p65 interacts with Keap1 to repress the Nrf2-ARE pathway. Cellular Signalling. 2011;23(5):883–92.

[41] Liu GH, Qu J, Shen X. NF-kappaB/p65 antagonizes Nrf2-ARE pathway by depriving CBP from Nrf2 and facilitating recruitment of HDAC3 to MafK. Biochim et Biophysica Acta. 2008;1783(5):713-27.

[42] Gloire G, Legrand-Poels S, Piette J. NF-kappaB activation by reactive oxygen species: fifteen years later. Biochemical Pharmacology. 2006;72(11):1493–505.

[43] Hwang YJ, Lee EW, Song J, Kim HR, Jun YC, Hwang KA. MafK positively regulates NF-kappaB activity by enhancing CBP-mediated p65 acetylation. Scientific Reports. 2013;3:3242.

[44] Rushworth SA, Zaitseva L, Murray MY, Shah NM, Bowles KM, MacEwan DJ. The high Nrf2 expression in human acute myeloid leukemia is driven by NF-kappaB and underlies its chemo-resistance. Blood. 2012;120(26):5188–98.

[45] Cuadrado A, Martin-Moldes Z, Ye J, Lastres-Becker I. Transcription factors NRF2 and NF-kappaB are coordinated effectors of the Rho family, GTP-binding protein RAC1 during inflammation. The Journal of Biological Chemistry. 2014;289(22): 15244–58.

[46] Lau A, Wang XJ, Zhao F, Villeneuve NF, Wu T, Jiang T, et al. A noncanonical mechanism of Nrf2 activation by autophagy deficiency: direct interaction between Keap1 and p62. Molecular and Cellular Biology. 2010;30(13):3275–85.

[47] Komatsu M, Kurokawa H, Waguri S, Taguchi K, Kobayashi A, Ichimura Y, et al. The selective autophagy substrate p62 activates the stress responsive transcription factor Nrf2 through inactivation of Keap1. Nature Cell Biology. 2010;12(3):213–23.

[48] Wooten MW, Geetha T, Seibenhener ML, Babu JR, Diaz-Meco MT, Moscat J. The p62 scaffold regulates nerve growth factor-induced NF-kappaB activation by influencing TRAF6 polyubiquitination. The Journal of Biological Chemistry. 2005;280(42):35625-9.

[49] Rada P, Rojo AI, Chowdhry S, McMahon M, Hayes JD, Cuadrado A. SCF/{beta}-TrCP promotes glycogen synthase kinase 3-dependent degradation of the Nrf2 transcription factor in a Keap1-independent manner. Molecular and Cellular Biology. 2011;31(6): 1121–33.

[50] Park SH, Park-Min KH, Chen J, Hu X, Ivashkiv LB. Tumor necrosis factor induces GSK3 kinase-mediated cross-tolerance to endotoxin in macrophages. Nature Immunology. 2011;12(7):607–15.

[51] Kang KW, Cho MK, Lee CH, Kim SG. Activation of phosphatidylinositol 3-kinase and Akt by tert-butylhydroquinone is responsible for antioxidant response element-mediated rGSTA2 induction in H4IIE cells. Molecular Pharmacology. 2001;59(5):1147–56.

[52] Li MH, Cha YN, Surh YJ. Peroxynitrite induces HO-1 expression via PI3K/Akt-dependent activation of NF-E2-related factor 2 in PC12 cells. Free Radical Biology and Medicine. 2006;41(7):1079–91.

[53] Hayes JD, Chowdhry S, Dinkova-Kostova AT, Sutherland C. Dual regulation of transcription factor Nrf2 by Keap1 and by the combined actions of beta-TrCP and GSK-3. Biochemical Society Transactions. 2015;43(4):611–20.

[54] Pitha-Rowe I, Liby K, Royce D, Sporn M. Synthetic triterpenoids attenuate cytotoxic retinal injury: cross-talk between Nrf2 and PI3K/AKT signaling through inhibition of the lipid phosphatase PTEN. Investigative Ophthalmology and Visual Science. 2009;50(11):5339–47.

[55] Rojo AI, Rada P, Mendiola M, Ortega-Molina A, Wojdyla K, Rogowska-Wrzesinska A, et al. The PTEN/NRF2 axis promotes human carcinogenesis. Antioxidants and Redox Signaling. 2014;21(18):2498–514.

[56] Taguchi K, Hirano I, Itoh T, Tanaka M, Miyajima A, Suzuki A, et al. Nrf2 enhances cholangiocyte expansion in PTEN-deficient livers. Molecular and Cellular Biology. 2014;34(5):900–13.

[57] Darvekar SR, Elvenes J, Brenne HB, Johansen T, Sjottem E. SPBP is a sulforaphane induced transcriptional coactivator of NRF2 regulating expression of the autophagy receptor p62/SQSTM1. PLoS One. 2014;9(1):e85262.

[58] Lau A, Zheng Y, Tao S, Wang H, Whitman SA, White E, et al. Arsenic inhibits autophagic flux, activating the Nrf2-Keap1 pathway in a p62-dependent manner. Molecular and Cellular Biology. 2013;33(12):2436–46.

[59] Yu R, Lei W, Mandlekar S, Weber MJ, Der CJ, Wu J, et al. Role of a mitogen-activated protein kinase pathway in the induction of phase II detoxifying enzymes by chemicals. The Journal of Biological Chemistry. 1999;274(39):27545–52.

[60] Yu R, Mandlekar S, Lei W, Fahl WE, Tan TH, Kong AN. p38 mitogen-activated protein kinase negatively regulates the induction of phase II drug-metabolizing enzymes that detoxify carcinogens. The Journal of Biological Chemistry. 2000;275(4):2322–7.

[61] Cullinan SB, Zhang D, Hannink M, Arvisais E, Kaufman RJ, Diehl JA. Nrf2 is a direct PERK substrate and effector of PERK-dependent cell survival. Molecular and Cellular Biology. 2003;23(20):7198–209.

[62] 6262. Huang HC, Nguyen T, Pickett CB. Regulation of the antioxidant response element by protein kinase C-mediated phosphorylation of NF-E2-related factor 2. Proceedings of the National Academy of Sciences of the United States of America. 2000;97(23):12475–80.

[63] Huang HC, Nguyen T, Pickett CB. Phosphorylation of Nrf2 at Ser-40 by protein kinase C regulates antioxidant response element-mediated transcription. The Journal of Biological Chemistry. 2002;277(45):42769–74.

[64] Niture SK, Jain AK, Jaiswal AK. Antioxidant-induced modification of INrf2 cysteine 151 and PKC-delta-mediated phosphorylation of Nrf2 serine 40 are both required for stabilization and nuclear translocation of Nrf2 and increased drug resistance. Journal of Cell Science. 2009;122(Pt 24):4452–64.

[65] Ahn YH, Hwang Y, Liu H, Wang XJ, Zhang Y, Stephenson KK, et al. Electrophilic tuning of the chemoprotective natural product sulforaphane. Proceedings of the National Academy of Sciences of the United States of America. 2010;107(21):9590-5.

[66] de Zeeuw D, Akizawa T, Audhya P, Bakris GL, Chin M, Christ-Schmidt H, et al. Bardoxolone methyl in type 2 diabetes and stage 4 chronic kidney disease. The New England Journal of Medicine. 2013;369(26):2492–503.

[67] DeNicola GM, Karreth FA, Humpton TJ, Gopinathan A, Wei C, Frese K, et al. Oncogene-induced Nrf2 transcription promotes ROS detoxification and tumorigenesis. Nature. 2011;475(7354):106–9.

[68] Sayin VI, Ibrahim MX, Larsson E, Nilsson JA, Lindahl P, Bergo MO. Antioxidants accelerate lung cancer progression in mice. Science Translational Medicine. 2014;6(221): 221ra15.

[69] Le Gal K, Ibrahim MX, Wiel C, Sayin VI, Akula MK, Karlsson C, et al. Antioxidants can increase melanoma metastasis in mice. Science Translational Medicine. 2015;7(308): 308re8.

[70] Piskounova E, Agathocleous M, Murphy MM, Hu Z, Huddlestun SE, Zhao Z, et al. Oxidative stress inhibits distant metastasis by human melanoma cells. Nature. 2015;527(7577):186–91.

[71] Ristow M, Zarse K, Oberbach A, Kloting N, Birringer M, Kiehntopf M, et al. Antioxidants prevent health-promoting effects of physical exercise in humans. Proceedings of the National Academy of Sciences of the United States of America. 2009;106(21):8665–70.

Chapter 4

Feed-Forward and Feed-Back Circuits of the NRF2/AP-1 Composite Pathway

Peter V. Zolotukhin and Anna A. Belanova

Additional information is available at the end of the chapter

http://dx.doi.org/10.5772/65216

Abstract

Being the central regulator of oxidative status of the cell, NRF2 must be regulated so that its activity can be rapidly and strongly induced when needed and quickly suppressed when not. Moreover, for the cell, NRF2 means much more than just antioxidant defense. Numerous general functions rely on NRF2 and related factors. All this implies that the NRF2 pathway has peculiar and powerful mechanisms of control of its activity. To a great extent, these mechanisms are based on feed-forward and feed-back circuits. These circuits, more than a dozen, are in the focus of this chapter.

Keywords: feed-back regulation, feed-forward regulation, SQSTM1, SESN2, thioredoxin, NF-kappaB, GSTP1, NAPDH oxidase, NRF1, NFE2L1, NRF3, NFE2L3, BACH1, 26S proteasome, MIR-144, mitochondrial biogenesis, truncated NRF2, NRF2, NFE2L2, AP-1

1. Introduction

According to the study by Malhotra et al. [1], Nrf2 controls 1055 protein-coding genes in mice. Although no similar study has ever been performed in humans and given that most routinely studied Nrf2 targets are the same in humans and mice, it is a reasonable assumption that roughly the same number of protein-coding genes is regulated by NRF2 in our species, accounting for astonishing ~5% of all our protein-coding genes. This raises a question: how does NRF2, being a well-known stand-by inducible transcription factor curbed in cytoplasm by KEAP1, fit into cellular context with so many target genes? Among those numerous genes, only a portion codes for immediate antioxidants and detoxifying enzymes. It is the more intriguing since central proteasomal [2], autophagic [3], general signaling [4, 5], and, moreover, cell proliferation, cell cycle and survival regulation factors [1] are in fact NRF2 targets.

Although 645 of the 1055 Nrf2 target genes were found to have basal type of expression regulation by this factor [1], Nrf2 is still required to be present in the nucleus in significant amounts to drive their expression. Notably, most of these genes are not at all related to antioxidant and xenobiotic defense systems, i.e., they are essential for proliferation and basic functioning of the cell. Thus, basal expression driven by NRF2 is important for the cell, but to what extent? One should bear in mind that NRF2 knockouts are nonlethal and generally appear normal in stable laboratory conditions [6], yet they are susceptible to numerous diseases [7–9], especially when the housing conditions are suboptimal [6, 10–12]. This confirms that basal NRF2 functioning, although not vital, is critical for the cell and organism even when no pro-oxidant or xenobiotic exposure is present.

It implies that some cellular mechanisms serve to provide the cell with at least minimal necessary NRF2 activity. On the other hand, it is inherently important for an inducible antioxidant and xenobiotic defense pathway to act as quickly as possible, thus signal amplification is required whenever the initial stimulus is applied. These two distinct prerequisites for normal cell functioning in ever-changing environment appear to be resolved in principle similarly: there are feed-forward circuits that maintain steady-state level of the NRF2 activity and make signal amplification possible when fast cellular reactions are required.

Conversely, too much NRF2 activity is no better than its absence: Keap1-null mice demonstrate postnatal lethality [13, 14], an effect coinciding with nuclear Nrf2 localization [14] and reversed by Nrf2 downregulation [13]. In this case, several feed-back circuits protect the cell and organism from obsessive NRF2 activation.

Most of the feed-forward and feed-back circuits of the NRF2 pathway also involve the closely related transcription factor activator protein 1 (AP-1) represented by homodimers of Jun proteins (JUN, JUNB, JUND) or heterodimers of Fos (FOS, FOSB, FOSL1, FOSL2) and Jun proteins. There are also other dimers of Jun/Fos proteins—e.g. with Atf-proteins [15–17]—still referred to as AP-1 complexes (although either Jun or Fos proteins are absent), and oligomers of Jun/Fos proteins with other proteins. NRF2, in line with NRF1, NRF3 and AP-1 proteins (including Atf proteins) are all basic leucine zipper (bZip) transcription factors of similar structure [18–20]. Functional roles and regulation of Nrf- and AP-1-proteins also significantly overlap [21–24], and, what is even more, these two groups of proteins regulate each other at several levels, including transcription control. It should be especially noted that NRF2 binding site ARE and AP-1 binding site TRE often overlap with AP-1 being embedded into ARE [25, 26]. Having such tight relations, NRF2 and AP-1 obviously form a composite NRF2/AP-1 (or NFE2L2/AP-1) pathway, with both factors contributing much to their own functioning via shared feed-forward and feed-back circuits.

All feed-forward and feed-back circuits are represented, but not outlined or highlighted, in oxidative status interactome map (OSIM) created in our laboratory [26]. This chapter, in contrast, emphasizes on functioning of these circuits and their peculiarities.

2. Feed-forward circuits

2.1. The SESN2/SQSTM1 circuit

In 2010, Copple et al., and in 2011, Bui and Shin independently demonstrated that the autophagosomal adaptor protein SQSTM1 (also known as p62) is capable of activating the NRF2 pathway without oxidative modification of KEAP1 protein [27, 28]. This route of the NRF2 pathway activation is one of many known for today. And by the time of this important discovery, it had already been known that *SQSTM1* itself is a target gene of NRF2 [29]. Thus, SQSTM1 was found to form a feed-forward circuit capable of noncanonical activation of NRF2. The mode of action of this factor is in line with its primary function—it merely targets KEAP1 for autophagosomal degradation, and this process takes place in the cytoplasmic complex of KEAP1 and NRF2 [27, 28]. Copple et al., of immediate relevance to this chapter, also showed that Sqstm1-dependent activation of Nrf2 is responsible for ~50% of basal expression of classical Nrf2/Ap-1 targets *Nqo1*, *Gclc* and classical Nrf2 target *Hmox1* [27].

Later on, Bae et al. [30] revealed that, in mice, Sqstm1 actually promotes Keap1 degradation when Sesn2 is available. A year before, Shin et al. found that *Sesn2/SESN2* is also an Nrf2/NRF2 target (**Figure 1**) [31].

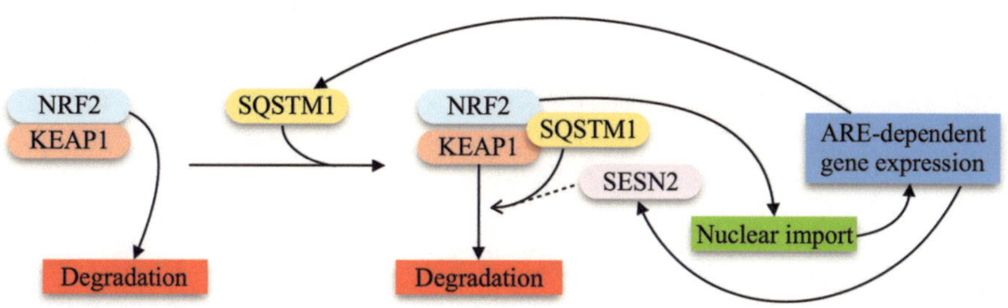

Figure 1. The SESN2/SQSTM1 circuit. Dashed line marks putative interaction.

To our knowledge, to date, there are no direct evidences that the same SESN2/SQSTM1 system is required for KEAP1 degradation. However, considering evolutionary significance of autophagy and, consequently, SQSTM1, it is highly possible that this holds true for humans as it does for mice. It is also known that SESN2 works together with SQSTM1 in mitophagy, an instance of autophagy [32]. As such, we ourselves have recently tested this hypothesis in our laboratory using the RNA interference approach. We found that in basal conditions *SESN2* knockdown caused significant changes in expression of *GCLC*, *HMOX1* and *SQSTM1* (all NRF2 targets tested in the study) [33]. Indirect observations emanating from this study also suggest that actually *BACH1* expression is also affected, and BACH1 forms a peculiar feed-back circuit which will be discussed below. In the same study, in pro-oxidant exposure conditions (modeled by culturing HeLa cells in medium containing 400 μM hydrogen peroxide for 24 h after the substance injection; 400 μM hydrogen peroxide was found to be sublethal for HeLa cells in our previous study [34]), on the contrary, *SESN2* knockdown caused less pronounced effect on expression of the same set of genes—only

HMOX1 and *SQSTM1* had significant changes in expression [33]. To the moment of preparation of this chapter, these data are in publication progress.

All together, the existing data clearly show that SESN2/SQSTM1 feed-forward circuit supports basal expression of NRF2/AP-1 targets and to some extent fortifies inducible reaction of this pathway.

2.2. The TXN feed-forward circuit

Thioredoxin 1, TXN, has long been known as a critical factor of activation of AP-1 transcription factor [35–37]. This TXN role has an intermediate factor—a DNA repair enzyme and a transcription factor DNA-binding promoter APEX1 [35, 36]. Interestingly, this protein is itself activated by some stimuli that are within primary focus of the NRF2/AP-1 pathway, including pro-oxidants and ionizing radiation [38], but excluding UV radiation [38], as is TXN nuclear import [39, 40].

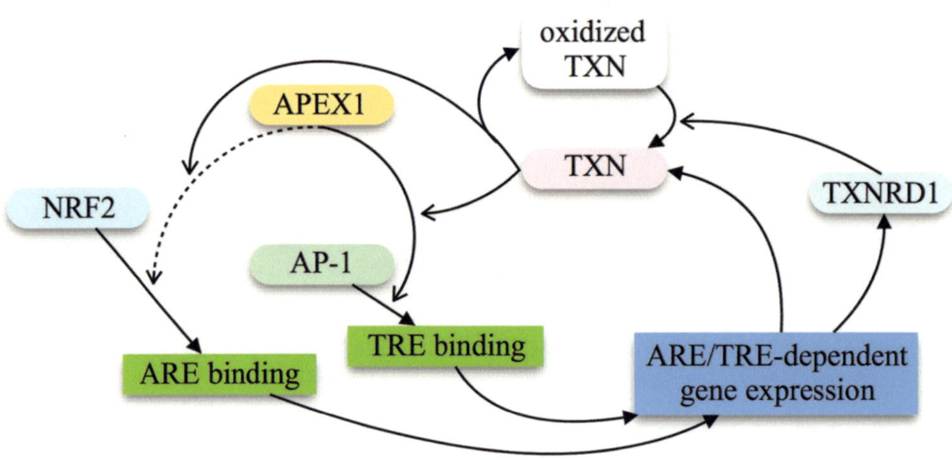

Figure 2. The TXN feed-forward circuit. Dashed line marks putative interaction.

Years later, Iwasaki et al. found that APEX1 actually facilitates DNA binding to expression of target genes of NRF2 [23] in addition to those of AP-1, NF-kappaB and HIF1A [41]. The same was also confirmed by Shan et al. [42]. As APEX1 reduction by TXN is obligate to its function [37], TXN is a player in NRF2-driven expression, at least in some instances. In this sense, it should be mentioned that the fact that TXN forces NRF2 binding to DNA may be due to direct heterodimerization of Juns (e.g., JUND) with NRF2 [23, 43, 44], as it exactly is in case of *FTH1*, which was the gene of interest in the study by Iwasaki et al. This raises the possibility that at least in some cases TXN stimulates NRF2 binding to DNA when AP-1 partners are present. Nevertheless, TXN stimulates the NRF2/AP-1 pathway activation.

At the same time, it has already been known for a long time that *TXN* expression is driven by antioxidant responsive element [45, 46], and Kim et al. directly showed that thioredoxin forces its own expression [22].

TXN functioning as a transcription factor DNA-binding stimulator is conferred by its participation in disulfide-dithiol exchange reactions with APEX1. Once APEX1 has reduced itself in expense of reduced TXN, TXN has to be reduced. There are two major reductases of TXN: TXNRD1 [47] and TXNRD3 [48]. *TXNRD1* is in fact an NRF2 target itself [49]. This makes the TXN feed-forward circuit self-sufficient (**Figure 2**). We found no evidences that *TXNRD3* is an NRF2 target in human, although this is still plausible since this gene/protein is merely poorly studied.

Thioredoxin feed-forward circuit is notable in one more sense: as TRE is often embedded in ARE, one should always consider that an ARE-containing gene may in fact be regulated by AP-1 binding to embedded TRE. This is exactly the case of thioredoxin 1: in their work, Kim et al. found that Jun and Fos overexpression decreased *TXN* expression in K562 cells [22]. In our work, we have also observed negative interactions between AP-1 and NRF2 working on the same ARE (with embedded TRE) [34], and this mechanism appears to be cell-context dependent, since NRF2 and AP-1 can have both positive and negative effects on the same genes in different cells. This will be discussed in detail in the next section.

Interestingly, TXN feed-forward circuit appears to involve a mechanism of adjustment, which is based on differential control of *TXN* transcripts. Among NRF2 targets, *BACH1* was the first to be described as having individual control of transcripts [50]. Later, we tested whether *TXN* transcript variants are differentially expressed when the NRF2 pathway gets activated, and so it was [51, 52]. This renders the TXN feed-forward circuit easily adjustable by the cell since only one of the transcripts is NRF2/AP-1 dependent.

The existence of this feed-forward circuit was proven experimentally. However, there are at least two more circuits in the NRF2/AP-1 pathway involving TXN, NRF2 and AP-1. One of them is a mixed feed-forward/feed-back circuit depending on cellular context (discussed in next section), whereas another is a feed-back circuit (discussed below with other feed-back circuits).

2.3. The AP-1/TXN/NRF2 bidirectional circuit

As *TXN* is an NRF2 target, and TXN is required for AP-1 activation, it forms one more complex circuit. This circuit cannot be considered as either purely feed forward or feed-back. In a set of studies, it was shown that JUN, being a part of AP-1, controls expression of *NRF2* [34, 53, 54]. The complexity is that JUN can both activate and suppress the *NRF2* expression depending on the cells analyzed: DeNicola et al. [53] demonstrated that, in murine MEF K-RasG12D cells, Jun transactivates *Nrf2*, whereas Cho et al. [54] and our laboratory [34] showed that JUN has negative influence on *NRF2* expression.

Thus, in the cells that are featured by JUN activating *NRF2*, JUN stimulates expression of *NRF2*, the latter activates expression of *TXN*, which, in turn, induces JUN DNA binding. In this case, this circuit acts as feed-forward for both NRF2 and AP-1 (**Figure 3A**).

In contrast, in cells whose JUN suppresses *NRF2*, less TXN is expressed and that is less favorable for AP-1 DNA binding, thus releasing suppression from NRF2 expression. Consequently, in this case, the circuit acts as feed-back for AP-1, whereas feed-forward, to some extent, for NRF2 (**Figure 3B**). It appears that this system favors NRF2 functioning in basal conditions, but acts to suppress it once NRF2 is activated. We ourselves observed the latter situation [34] but have not yet tested the former.

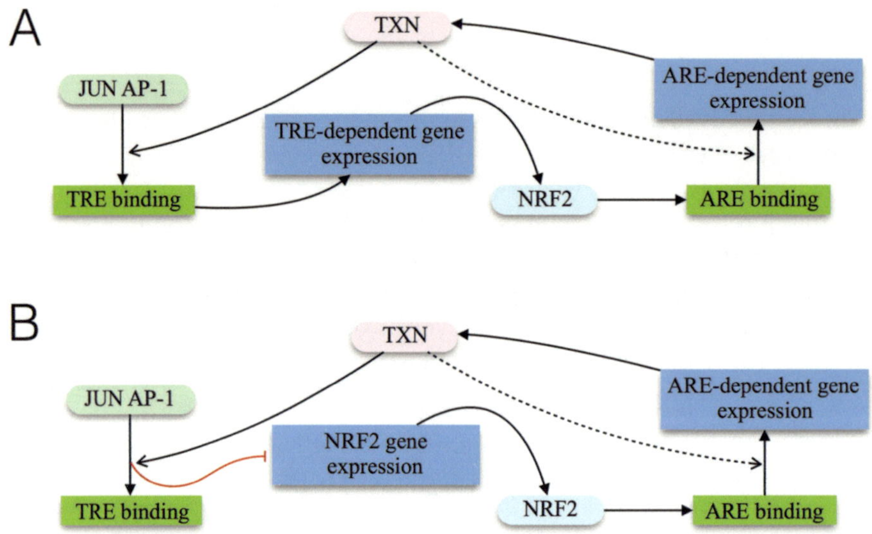

Figure 3. The AP-1/TXN/NRF2 bidirectional circuit. Dashed line marks putative interaction. (A) depicts the situation when cell signaling background determines the system of interaction between NRF2 and AP-1 to act as a feed-forward circuit. On the other hand, (B) illustrates how these interactions form a feed-back circuit in other cell types or cellular conditions.

How is this possible that this circuit differs this much in various cells? The simplest explanation we would suggest is, different cell lines express different amounts of Fos proteins—FOS, FOSB, FOSL1, FOSL2, as well as other Jun proteins—JUNB, JUND [22, 35, 55]. All these proteins significantly differ in their effects on JUN. Moreover, how all Juns perform depends on the presence of JUN dimerizing protein JDP2 capable of deactivating Juns by dimerizing them and by epigenetically silencing the *JUN* gene [56]. Thus, immediate protein partners of JUN and its powerful regulator JDP2 seem to determine whether JUN will activate or suppress the NRF2 expression in a given cell line. However, this is to be tested in upcoming studies.

2.4. The GSTP1 feed-forward circuit

One of the well-known NRF2 targets, *GSTP1* [57, 58], has recently been found to form a novel feed-forward circuit of the NRF2/AP-1 pathway. Carvalho et al. have just recently revealed that, in mice, Gstp1 is capable of S-glutathionylation of Keap1 disrupting the interactions between Keap1 and Nrf2 (**Figure 4**) [59].

Although this has not been observed in humans to date, it is plausible that the same circuit is also characteristic of the human cell. It is also tempting to suggest that other glutathione-S-transferases may have the same function. At least *GSTA1* [60] and *GSTA4* [61] expression is, meanwhile, subject to NRF2 regulation.

http://dx.doi.org/10.5772/65216

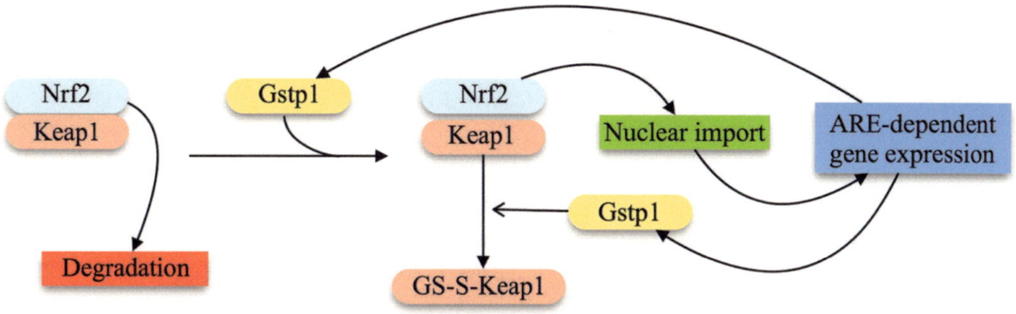

Figure 4. The GSTP1 feed-forward circuit. GS-S-Keap1—S-glutathionylated Keap1.

Of importance, *GSTP1* is activated by both AP-1 and NRF2. Thus, whenever AP-1 is activated, NRF2 is released of KEAP1. This may be a mechanism of protective coupling of cellular events because AP-1 has much more profound effects on the cellular physiology than NRF2 does.

2.5. The NOX4 feed-forward circuit

Among peculiar NRF2 targets is NOX4 [62, 63]. This protein is a part of one of NADPH oxidase multiprotein complexes. The inherent NOX4 function is production of superoxide anion, similarly to other Noxes. Two features of this enzyme with respect to other enzymes of the family are that (1) NOX4 is strongly expressed in kidney where it is suggested to serve as an oxygen sensor for controlling the erythropoietin production [64, 65], and (2) NOX4 is localized to endoplasmic reticulum [66] and nucleus [65] in addition to cell membrane.

Barring NOX5 [67], Noxes require partners for functioning. In case of NOX4, the obligate partner is only one—the CYBA protein (widely known as p22phox) [68]. Thus, the NOX4 feed-forward circuit would require CYBA for functioning, and *CYBA* is not an NRF2 target.

However, *CYBA* is in fact an NF-kappaB and AP-1 target [69]. Thus, the composite NRF2/AP-1 pathway controls both enzymes necessary for the NOX4 feed-forward circuit (**Figure 5**).

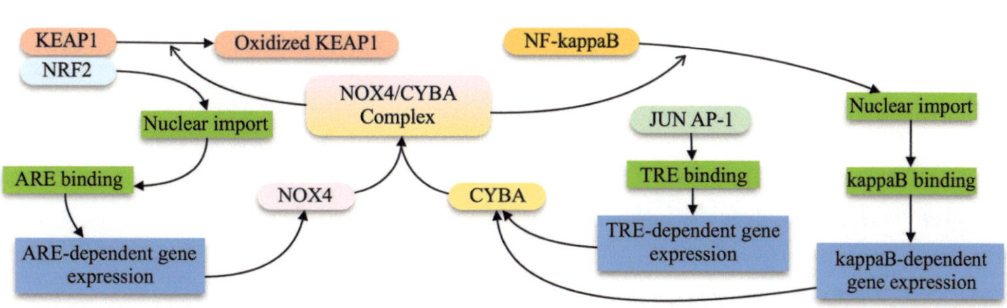

Figure 5. The NOX4 feed-forward circuit.

Additionally, NF-kappaB is activated by pro-oxidants [40, 70–72], and DNA binding of both AP-1 and NF-kappaB is stimulated by TXN [40, 70–73].

3. Feed-back circuits

As seen from data above, the cellular feed-forward circuits of the NRF2/AP-1 pathway are numerous and versatile providing the cell with required steady-state levels of the proteins, and signal amplification upon induction of the pathway. However, as already mentioned, hyperactive NRF2 and AP-1 are a death threat to the organism. If NRF2 causes disturbances of tissues functioning, AP-1 is a pro-oncogenic transcription factor (JUN and FOS proteins are also, and in some areas—ordinarily, known as c-jun and c-fos proto-oncogenes) [74, 75]. In addition to that, NRF2 controls autophagy through SQSTM1, and also transactivates several ABC-transporters, including *ABCB1* (*MDR1*), *ABCG2* (*BCRP*), *ABCC2* (*MRP2*) and *ABCC3* (*MRP3*) [76]—these are all major factors of carcinogenesis and cancer progression. Please note that although autophagy suppresses malignization, once tumor turns malignant, autophagy becomes a cancer-promoting metabolic and drug resistance-conferring trait [77].

Thus, NRF2 and AP-1 activity should be tightly controlled once activated. Several feed-back circuits allow for that. Two of them involve factors participating in feed-forward circuits. These circuits are the first to be discussed.

3.1. The TXN feed-back circuit

TXN localizes to cytoplasm and migrates to nucleus upon stimulation. However, the cytoplasmic pool of thioredoxin 1 provides the cell with antioxidant capacity. Thus, cytoplasmic TXN inhibits pro-oxidants-induced NF-kappaB and NRF2 nuclear import [72].

Consequently, as soon as the NRF2/AP-1 pathway activation produces substantial amounts of thioredoxin 1, TXN stabilizes KEAP1 and promotes NRF2 sequestration and degradation (**Figure 6**). This works wherever KEAP1 and TXN appear, as will be discussed later.

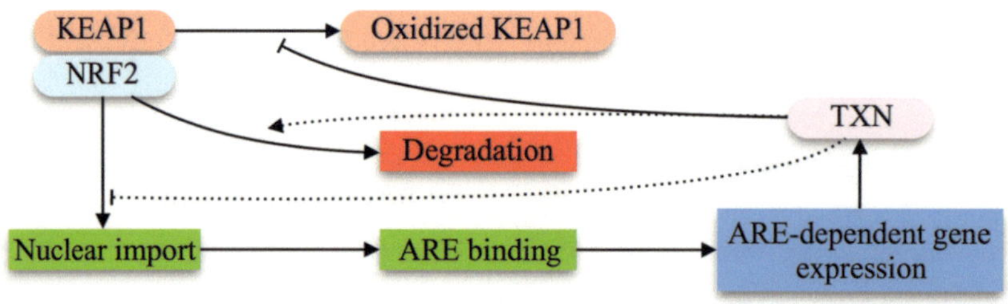

Figure 6. The TXN feed-back circuit. Dotted line marks an indirect effect.

3.2. The NRF3/NOX4 feed-back circuit

One of the least studied, if not the worst studied, bZip proteins is the third homolog of NRF2—NRF3. Very little is known about this transcription factor except the fact that it usually antagonizes NRF1 and NRF2 [22, 78, 79]. Even less is known on stimuli that activate NRF3. NRF3 knockout is known not to change the phenotype to any notable extent [80].

However, Zhang et al. and Pepe et al. revealed that NRF3 is activated by ER stress inducers [81, 82]. Once activated, NRF3, predictably, suppresses ordinary NRF2 target genes [82]. Surprisingly, in contrast to these genes, *NOX4* was activated by NRF3 [82]. As NOX4 is an enzyme localized to endoplasmic reticulum and producing superoxide anion, it is capable of oxidative modification of inositol 1,4,5-trisphosphate receptors [83]. Once oxidized, these receptors cause calcium efflux from endoplasmic reticulum to cytosol. This event leads to endoplasmic reticulum stress [83].

As mentioned above, NOX4, for its functioning, requires an AP-1 target—CYBA. ER stress is not a leading, if at all, stimulus for AP-1 activation, yet ER stress causes ROS production (during the protein refolding period), and ROS activation is an acknowledged property of AP-1 [35]. Additionally, NRF3 heterodimerizes with AP-1 proteins FOS and FOSL1 (it was suggested by Zhang et al. that NRF3 thus suppresses TRE-containing NRF2 targets only [81]) and is suggested to heterodimerize with Jun proteins (**Figure 7**) [79, 84].

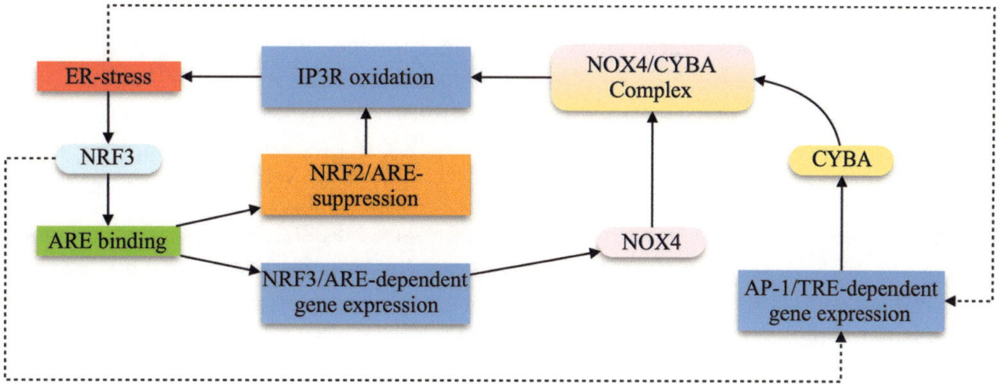

Figure 7. The NRF3/NOX4 feed-back circuit. Dashed line marks putative interaction.

The fact that NRF3 activates *NOX4* also suggests that the NOX4 feed-forward circuit may equally work for both NRF2 and AP-1.

3.3. The KEAP1 feed-back circuit

KEAP1 is the immediate antagonist of NFR2 as it sequesters it and targets it for proteasomal degradation. In a complex experiment, Lee et al. demonstrated that murine *Keap1* is another Nrf2 target [85]. Unfortunately, no direct evidences that the same is true for humans are known to the authors to date of preparation of this chapter. This is an extremely important issue to be addressed in future studies. Nevertheless, at least in mice, this circuit is active and facilitates fast Nrf2 pathway shutdown as soon as the oxidative status of the cell is normalized (**Figure 8**).

Interestingly, KEAP1 can be imported into the nucleus by KPNA6 [46, 60, 86]. In the nucleus, KEAP1 acts exactly as in the cytoplasm—it binds NFE2L2 and induces its polyubiquitination thus targeting it for degradation [46, 85].

It means that GSK3B activity providing NRF2 nuclear export [87–89] is not even required for KEAP1 to force NRF2 degradation. However, KEAP1 nuclear import is a tightly controlled process. An example of the chain of events causing KEAP1 nuclear import is presented in the next section.

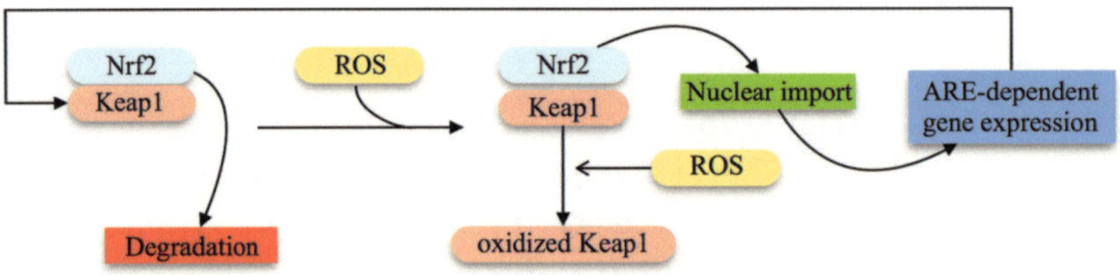

Figure 8. The KEAP1 feed-back circuit.

3.4. The NF-kappaB/KEAP1 feed-back circuit

KEAP1 nuclear import stimulation is well described for the case of NF-kappaB activation. Generally, the NF-kappaB and NRF2 pathways antagonize in several ways: NF-kappaB protein RELA (p65) promotes HDAC3 interaction with CREBBP or MAFK thus causing local hypoacetylation surrounding ARE [90]; KEAP1 promotes IKK degradation thus suppressing NF-kappaB release from IKBs [91]; and, in turn, RELA promotes KEAP1 nuclear import [46].

Thus, whenever NF-kappaB is activated, it suppresses NFE2L2 by two mechanisms: by inducing its nuclear sequestration by KEAP1 and by causing transcription-suppressing epigenomic modification of the NRF2-dependent loci. In this sense, the NF-kappaB/KEAP1 circuit presented here is feed-back for the NRF2/AP-1 pathway yet feed-forward for the NF-kappaB pathway, because less KEAP1 causes less IKK degradation consequently promoting NF-kappaB nuclear import. This feed-forward circuit is probably disrupted by TXN, which is NRF2-dependent (**Figure 9**).

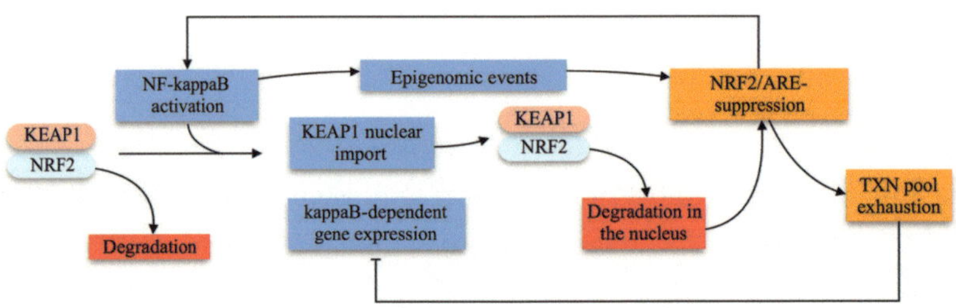

Figure 9. The NF-kappaB/KEAP1 feed-back circuit.

3.5. The BACH1 feed-back circuit

In contrast to the preferentially cytoplasmic NRF2 inhibitor KEAP1, BACH1 is its nuclear antagonist. Just as NRF2, BACH1 belongs to the bZip family, and moreover, to cap'n'collar

http://dx.doi.org/10.5772/65216

(CNC) sub-family. Thus, BACH1 and NRF2 act in a very similar manner—by heterodimerizing with small Maf proteins [86, 92–94]. There is, however, one striking difference between the two transcription factors: BACH1 acts only on clustered-ARE genes [49, 93], like *HMOX1* [93, 95] and *NQO1* [96].

As it was discussed earlier, different NRF2 targets may response distinctly to the pathway stimulation or release of pathway suppression. This actually may be caused by BACH1. For example, *TXNRD1* is proven to be BACH1-independent [49], while *HMOX1* has quadruple ARE [93] being subject to BACH1-induced suppression, and *NQO1* has a double ARE still being somewhat suppressed by BACH1 [96]. It appears that BACH1 functioning allows the cell to discriminate between the NRF2 targets. It is achieved by a relatively simple mechanism: if NRF2 activity is induced by whatever mechanism or stimulus except for pro-oxidant exposure, BACH1 does not allow NRF2 to act on the clustered-ARE genes. In contrast, when the cell is exposed to pro-oxidants of whatever nature, BACH1 is fast oxidized [19, 49, 86, 93, 96, 97], rapidly detaches from DNA [93, 94, 98] and readily degrades [93, 97].

At the same time, *BACH1* is itself an NRF2 target and an extremely interesting one. The *BACH1* locus codes for three transcript variants produced as the result of alternative transcription initiation. Only *BACH1* transcript variant 2 is NRF2-dependent: *BACH1* intronic +1411 nt ARE is functional and promotes transcription of the second intron of the gene [50]. The authors of this discovery, Jyrkkänen et al., proposed the existence of the BACH1 feed-back circuit along with their discovery of the functional intronic ARE [50].

The existence of the BACH1 feed-back circuit predisposes different expression dynamics of NRF2 targets depending on their ARE structure: once the NRF2 pathway normalizes oxidative status of the cell, newly synthesized BACH1 protein successfully outcompetes NRF2 on the clustered-ARE loci, while allowing it to act further on non-clustered ARE genes (**Figure 10**).

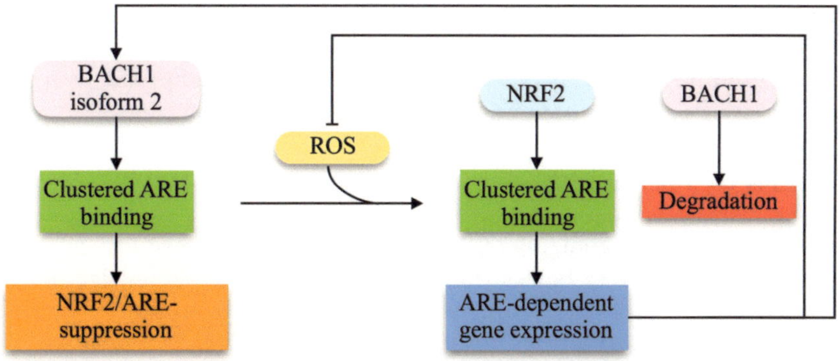

Figure 10. The BACH1 feed-back circuit.

In one of our studies, we appeared to observe these events after we released the JUN-imposed suppression of NRF2 in cells treated with hydrogen peroxide: in contrast non-clustered ARE genes (*FTH1, CBR3, SQSTM1*), *HMOX1* expression did not rise—probably because BACH1, being a non-clustered ARE gene, had increased expression in these those settings [34].

3.6. The 26S proteasome/NRF2 feed-back circuit

Once polyubiquitinated, NRF2 is degraded by 26S proteasome [86]. Proteasome inhibitors suppress the degradation of NRF2 and stimulate its nuclear import [99].

At the same time, as shown in mice, Nrf2 controls at least eight genes of proteasomal proteins: *Psma1*, *Psma4*, *Psmb3*, *Psmb5*, *Psmb6*, *Psmc1*, *Psmc3* and *Psmd14* (**Figure 11**) [2].

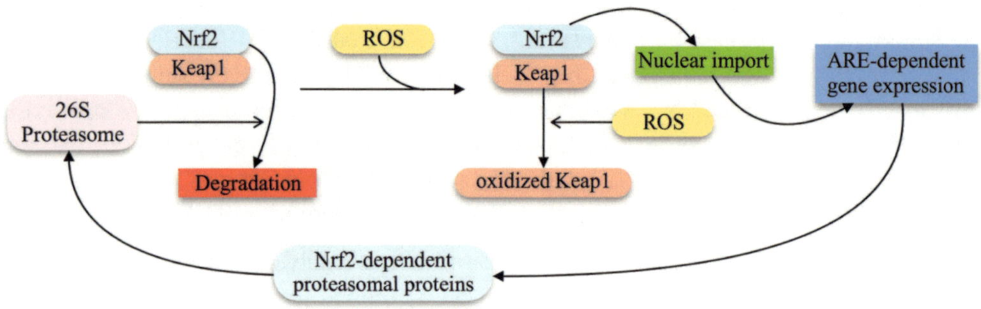

Figure 11. The 26S proteasome/NRF2 feed-back circuit.

To date of preparation of the chapter, no similar results obtained from studies on human cells have been published. Nevertheless, considering the biological significance of the 26S proteasome, it is not an unfair assumption that the same genes are probably controlled by NRF2 in humans.

Notably, even in human cells, AP-1 controls expression of interferon gamma [100], which, in turn, transactivates *PSME1* [101] and *PSME2* [102] subunit genes of the 26S proteasome.

Thus, NRF2 pathway activation facilitates NRF2 proteasomal degradation once KEAP1 is re-enabled to sequester NRF2. At least, this system works in mice, and probably in humans, too, especially since two *Psme* genes are under the NRF2/AP-1 pathway control.

3.7. The MIR-144 feed-back circuit

One of numerous NRF2-antagonizing miRNAs, *MIR-144*, is expressed under control of AP-1 [103]. This miRNA not merely blocks translation of NRF2, but rather degrades the *NRF2* mRNA [104, 105].

3.8. The mitochondrial bidirectional circuit

Being regulated by reactive oxygen species and signaling background of the cell, the NRF2/AP-1 pathway depends on mitochondrial function, including their vast signaling activity [106]. Surprisingly, the NRF2/AP-1 pathway has been found to modulate mitochondrial biogenesis by two mechanisms. The first one has direct experimental evidences: NRF2 is a transcriptional regulator of *NRF1*—nuclear respiratory factor (which is not to be confused with *NFE2L1*, an

NRF2 homolog, also known and referred to as *NRF1* in this chapter) [107]. NRF1, in turn, controls a battery of nuclear genes required for mitochondrial biogenesis and function [107]. The second mechanism has not been observed in direct experiments and is related to the fact that the NRF1 protein is also controlled by estrogen receptors—a distinct family of transcription factors [108]. However, it is well-known that estrogen receptors act in an intimate collaboration with AP-1 proteins [109, 110]. For example, FOS alone is required for expression of 37% of estrogen receptors target genes [111]. Even though the mitochondrial biogenesis-related targets of the estrogen receptors may not fall into the AP-1-dependent category, which is to be addressed in the future, there is another doubtless interaction between the estrogen receptors and the NRF2/AP-1 pathway. The already discussed NRF2/AP-1 target, *TXNRD1* is a strong promoter of transactivatory function of the estrogen receptors [111], thus the NRF2/AP-1 pathway positively regulates estrogen receptors function, while estrogen receptors facilitate mitochondria biogenesis.

Although mitochondrial biogenesis eventually leads to increased ROS generation, this circuit is unlikely to be purely or even mostly feed-forward. Rather, this circuit is bidirectional with strong feed-back action because the outer mitochondrial membrane carries the PGAM5 protein. PGAM5, in turn, tethers the NRF2-KEAP1 complexes to mitochondria and acts as a powerful suppressor of the NRF2-dependent expression (**Figure 12**) [106, 112].

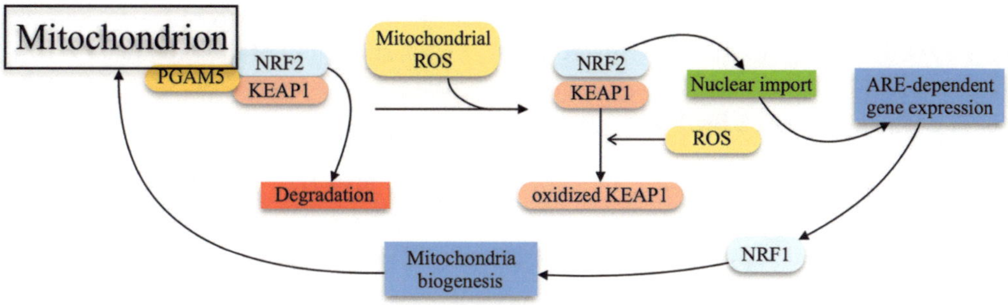

Figure 12. The mitochondrial bi-directional circuit. NRF1 here is nuclear respiratory factor 1.

Thus, the more mitochondria, the more KEAP1-NRF2 complexes are likely to be bound to them, leading to decreased availability of NRF2 to nucleus. Unfortunately, it is not known to date how PGAM5 expression is controlled with relation to mitochondria biogenesis—the protein is poorly studied.

It is worth noting that the PGAM5/KEAP1-NRF2 system is suggested to represent a stand-by alarm inducible by changes in mitochondrial function [106, 112]. PGAM5 also appears to stimulate mitophagy as a part of cellular defense against excessive reactive oxygen species [113]. It could be suggested that this function of PGAM5 is directly related to its sequestering of NRF2 with subsequent release of it upon mitochondrial dysfunction: NRF2 controls expression of *SQSTM1* and *SESN2*, two factors of autophagy/mitophagy already discussed above [32, 114].

3.9. The NRF2 bidirectional circuit

NRF2 expression is known to respond to stimuli that activate the NRF2/AP-1 pathway [34, 51, 115]. We and others observed that, for reasons still to be revealed, *NRF2* had a humped expression curve when different cells had been treated with various pro-oxidant substances [34, 115]. It is possible that this is related to the AP-1/TXN/NRF2 bidirectional circuit. It is tempting to suggest that this is caused by AP-1 components that are expressed with different dynamics upon stimulation.

For example, Siriani et al. demonstrated, that angiotensin II, over the same period of treatment, induced 300-fold change in expression of *FOS*, 500-fold change in expression of *FOSB*, 2-fold change in expression of *FOSL1* and *FOSL2*, 4-fold change in expression of *JUN*, 14-fold change in expression of *JUNB* and 2-fold change in expression of *JUND*. The same was also fair for the classical AP-1 inducer—TPA: it caused 194-fold change in expression of *FOS*, 65-fold change in expression of *FOSB*, 2-fold change in expression of *FOSL1*, 2-fold change in expression of *JUN* and *JUND*, and 9-fold change in expression of *JUNB* [116].

Since overflow inhibition is characteristic of AP-1 [117], and competition of bZip dimers is a must, it is possible that alterations in ratio of AP-1 components controlling *NRF2* expression lead to its humped dosage-expression curve.

This circuit cannot be considered purely feed-back, because before certain threshold, the stimulus activates *NRF2* expression. Yet, this character of expression of *NRF2* is not entirely of feed-forward nature, and this is discussed in the next, last, section of this chapter.

3.10. The truncated NRF2 feed-back circuit

NRF2 is cleaved into a truncated form named truncated NRF2, or tNRF2, by Ice family-like caspases [118]. Just as the full-length protein, tNRF2 enters nucleus, but there it antagonizes the normal NRF2 and suppresses ARE-containing loci [19, 118].

4. Conclusion

Although studying the NRF2/AP-1 pathway targets is an extremely complex task due to their plentitude, experimental evidences suggest that this relatively small, with respect to the regulatory proteins, pathway has enormous number of feed-forward and feed-back circuits, some of them being bidirectional.

Some evidences demonstrated directly that these circuits contribute to either fast activation or quenching of the pathway. For other circuits, no such testing has ever been performed. Nevertheless, the existing data unambiguously point out that these circuits are an essential feature of the NRF2/AP-1 pathway. These circuits must be accounted in all applications—starting from cell sensor-based pharmacological screening techniques [119, 120] and ending with therapeutical research and development [121].

Acknowledgements

Our studies and preparation of this chapter were supported by Russian Science Foundation Grant # 14-35-00051.

Author details

Peter V. Zolotukhin* and Anna A. Belanova

*Address all correspondence to: p.zolotukhin@icloud.com

Cell Physiology Laboratory, Evolution Corporate Group, Rostov-on-Don, Russia

References

[1] Malhotra D, Portales-Casamar E, Singh A, Srivastava S, Arenillas D, Happel C, Shyr C, Wakabayashi N, Kensler TW, Wasserman WW, Biswal S. Global mapping of binding sites for Nrf2 identifies novel targets in cell survival response through ChIP-Seq profiling and network analysis. Nucleic Acids Res. 2010;38(17):5718–5734. doi:10.1093/nar/gkq212.

[2] Kwak MK, Wakabayashi N, Greenlaw JL, Yamamoto M, Kensler TW. Antioxidants enhance mammalian proteasome expression through the Keap1-Nrf2 signaling pathway. Mol Cell Biol. 2003;23(23):8786–8794.

[3] Taguchi K, Fujikawa N, Komatsu M, Ishii T, Unno M, Akaike T, Motohashi H, Yamamoto M. Keap1 degradation by autophagy for the maintenance of redox homeostasis. Proc Natl Acad Sci USA. 2012;109(34):13561–13566. doi:10.1073/pnas.1121572109.

[4] Erttmann SF, Bast A, Seidel J, Breitbach K, Walther R, Steinmetz I. PGD2 and PGE2 regulate gene expression of Prx 6 in primary macrophages via Nrf2. Free Radic Biol Med. 2011;51(3):626–640. doi:10.1016/j.freeradbiomed.2011.05.022.

[5] Ho JN, Lee SB, Lee SS, Yoon SH, Kang GY, Hwang SG, Um HD. Phospholipase A2 activity of peroxiredoxin 6 promotes invasion and metastasis of lung cancer cells. Mol Cancer Ther. 2010;9(4):825–832. doi:10.1158/1535-7163.MCT-09-0904.

[6] Priestley JR, Kautenburg KE, Casati MC, Endres BT, Geurts AM, Lombard JH. The NRF2 knockout rat: a new animal model to study endothelial dysfunction, oxidant stress, and microvascular rarefaction. Am J Physiol Heart Circ Physiol. 2016;310(4):H478–487. doi:10.1152/ajpheart.00586.2015.

[7] Khor TO, Huang MT, Prawan A, Liu Y, Hao X, Yu S, Cheung WK, Chan JY, Reddy BS, Yang CS, Kong AN. Increased susceptibility of Nrf2 knockout mice to colitis-associated

colorectal cancer. Cancer Prev Res (Phila). 2008;1(3):187–191. doi: 10.1158/1940-6207.CAPR-08-0028.

[8] Beyer TA, Xu W, Teupser D, auf dem Keller U, Bugnon P, Hildt E, Thiery J, Kan YW, Werner S. Impaired liver regeneration in Nrf2 knockout mice: role of ROS-mediated insulin/IGF-1 resistance. EMBO J. 2008;27(1):212–223.

[9] Cheung KL, Lee JH, Khor TO, Wu TY, Li GX, Chan J, Yang CS, Kong AN. Nrf2 knockout enhances intestinal tumorigenesis in Apc(min/+) mice due to attenuation of anti-oxidative stress pathway while potentiates inflammation. Mol Carcinog. 2014;53(1):77–84. doi:10.1002/mc.21950.

[10] Furnari MA, Saw CL, Kong AN, Wagner GC. Altered behavioral development in Nrf2 knockout mice following early postnatal exposure to valproic acid. Brain Res Bull. 2014;109:132–142. doi:10.1016/j.brainresbull.2014.10.006.

[11] Chartoumpekis DV, Ziros PG, Zaravinos A, Iskrenova RP, Psyrogiannis AI, Kyriazopoulou VE, Sykiotis GP, Habeos IG. Hepatic gene expression profiling in Nrf2 knockout mice after long-term high-fat diet-induced obesity. Oxid Med Cell Longev. 2013;2013:340731. doi:10.1155/2013/340731.

[12] Becks L, Prince M, Burson H, Christophe C, Broadway M, Itoh K, Yamamoto M, Mathis M, Orchard E, Shi R et al. Aggressive mammary carcinoma progression in Nrf2 knockout mice treated with 7,12-dimethylbenz[a]anthracene. BMC Cancer. 2010;10:540. doi:10.1186/1471-2407-10-540.

[13] Wakabayashi N, Itoh K, Wakabayashi J, Motohashi H, Noda S, Takahashi S, Imakado S, Kotsuji T, Otsuka F, Roop DR et al. Keap1-null mutation leads to postnatal lethality due to constitutive Nrf2 activation. Nat Genet. 2003;35(3):238–245.

[14] Satoh T, Harada N, Hosoya T, Tohyama K, Yamamoto M, Itoh K. Keap1/Nrf2 system regulates neuronal survival as revealed through study of keap1 gene-knockout mice. Biochem Biophys Res Commun. 2009;380(2):298–302. doi:10.1016/j.bbrc.2009.01.063.

[15] Juilland M, Gonzalez M, Erdmann T, Banz Y, Jevnikar Z, Hailfinger S, Tzankov A, Grau M, Lenz G, Novak U et al. CARMA1- and MyD88-dependent activation of Jun/ATF-type AP-1 complexes is a hallmark of ABC diffuse large B-cell lymphomas. Blood. 2016;127(14):1780–1789. doi:10.1182/blood-2015-07-655647.

[16] Gozdecka M, Breitwieser W. The roles of ATF2 (activating transcription factor 2) in tumorigenesis. Biochem Soc Trans. 2012;40(1):230–234. doi:10.1042/BST20110630.

[17] Srivastava PK, Hull RP, Behmoaras J, Petretto E, Aitman TJ. JunD/AP1 regulatory network analysis during macrophage activation in a rat model of crescentic glomerulonephritis. BMC Syst Biol. 2013;7:93. doi:10.1186/1752-0509-7-93.

[18] Novotny V, Prieschl EE, Csonga R, Fabjani G, Baumruker T. Nrf1 in a complex with fosB, c-jun, junD and ATF2 forms the AP1 component at the TNF alpha promoter in stimulated mast cells. Nucleic Acids Res. 1998;26(23):5480–5485.

[19] Sykiotis GP, Bohmann D. Stress-activated cap'n'collar transcription factors in aging and human disease. Sci Signal. 2010;3(112):re3. doi:10.1126/scisignal.3112re3.

[20] Babu RL, Naveen Kumar M, Patil RH, Devaraju KS, Ramesh GT, Sharma SC. Effect of estrogen and tamoxifen on the expression pattern of AP-1 factors in MCF-7 cells: role of c-Jun, c-Fos, and Fra-1 in cell cycle regulation. Mol Cell Biochem. 2013;380(1–2):143–151. doi:10.1007/s11010-013-1667-x.

[21] Venugopal R, Jaiswal AK. Nrf2 and Nrf1 in association with Jun proteins regulate antioxidant response element-mediated expression and coordinated induction of genes encoding detoxifying enzymes. Oncogene. 1998;17(24):3145–3156.

[22] Kim YC, Yamaguchi Y, Kondo N, Masutani H, Yodoi J. Thioredoxin-dependent redox regulation of the antioxidant responsive element (ARE) in electrophile response. Oncogene.2003;22(12):1860–1865.

[23] Iwasaki K, Mackenzie EL, Hailemariam K, Sakamoto K, Tsuji Y. Hemin-mediated regulation of an antioxidant-responsive element of the human ferritin H gene and role of Ref-1 during erythroid differentiation of K562 cells. Mol Cell Biol. 2006;26(7):2845–2856.

[24] Xanthoudakis S, Miao G, Wang F, Pan YC, Curran T. Redox activation of Fos-Jun DNA binding activity is mediated by a DNA repair enzyme. EMBO J. 1992;11(9):3323–3335.

[25] Reichard JF, Petersen DR. Hepatic stellate cells lack AP-1 responsiveness to electrophiles and phorbol 12-myristate-13-acetate. Biochem Biophys Res Commun. 2004;322(3):842–853. doi:10.1016/j.bbrc.2004.07.180

[26] Zolotukhin P, Kozlova Y, Dovzhik A, Kovalenko K, Kutsyn K, Aleksandrova A, Shkurat T. Oxidative status interactome map: towards novel approaches in experiment planning, data analysis, diagnostics and therapy. Mol Biosyst. 2013;9(8):2085–2096. doi: 10.1039/c3mb70096h.

[27] Copple IM, Lister A, Obeng AD, Kitteringham NR, Jenkins RE, Layfield R, Foster BJ, Goldring CE, Park BK. Physical and functional interaction of sequestosome 1 with Keap1 regulates the Keap1-Nrf2 cell defense pathway. J Biol Chem. 2010;285(22):16782–16788. doi:10.1074/jbc.M109.096545.

[28] Bui CB, Shin J. Persistent expression of Nqo1 by p62-mediated Nrf2 activation facilitates p53-dependent mitotic catastrophe. Biochem Biophys Res Commun. 2011;412(2):347–352. doi:10.1016/j.bbrc.2011.07.101.

[29] Jain A, Lamark T, Sjøttem E, Larsen KB, Awuh JA, Øvervatn A, McMahon M, Hayes JD, Johansen T. p62/SQSTM1 is a target gene for transcription factor NRF2 and creates a positive feedback loop by inducing antioxidant response element-driven gene transcription. J Biol Chem. 2010;285(29):22576–22591. doi:10.1074/jbc.M110.118976.

[30] Bae SH, Sung SH, Oh SY, Lim JM, Lee SK, Park YN, Lee HE, Kang D, Rhee SG. Sestrins activate Nrf2 by promoting p62-dependent autophagic degradation of Keap1 and

prevent oxidative liver damage. Cell Metab. 2013;17(1):73–84. doi:10.1016/j.cmet.2012.12.002

[31] Shin BY, Jin SH, Cho IJ, Ki SH. Nrf2-ARE pathway regulates induction of Sestrin-2 expression. Free Radic Biol Med. 2012;53(4):834–841. doi:10.1016/j.freeradbiomed.2012.06.026.

[32] Kim MJ, Bae SH, Ryu JC et al. SESN2/sestrin2 suppresses sepsis by inducing mitophagy and inhibiting NLRP3 activation in macrophages. Autophagy. 2016;12(8):1272–1291. doi: 10.1080/15548627.2016.1183081.

[33] Belanova AA, Smirnov DS, Makarenko MS, Belousova MM, Zolotukhin PV. SESN2 knockdown deregulates expression of NFE2L2-dependent genes in HeLa cells in basal conditions and under pro-oxidant exposure. In press.

[34] Belanova AA, Smirnov DS, Makarenko MS, Kornienko IV, Kurbatov SV, Belousova MM, Aleksandrova AA, Kharchenko EY, Prazdnova EV, Zolotukhin PV. JUN knockdown reinforces expression of NFE2L2-dependent genes in hydrogen peroxide-treated HeLa cells. Indian J Exp Biol. 2015; publication in process, revision 1.

[35] Sun Y, Oberley LW. Redox regulation of transcriptional activators. Free Radic Biol Med. 1996;21(3):335–348.

[36] Abate C, Patel L, Rauscher FJ 3rd, Curran T. Redox regulation of fos and jun DNA-binding activity in vitro. Science. 1990;249(4973):1157–1161.

[37] Brandes N, Schmitt S, Jakob U. Thiol-based redox switches in eukaryotic proteins. Antioxid Redox Signal. 2009;11(5):997–1014. doi:10.1089/ARS.2008.2285.

[38] Ramana CV, Boldogh I, Izumi T, Mitra S. Activation of apurinic/apyrimidinic endonuclease in human cells by reactive oxygen species and its correlation with their adaptive response to genotoxicity of free radicals. Proc Natl Acad Sci USA. 1998;95(9):5061–5066.

[39] Wei SJ, Botero A, Hirota K, Bradbury CM, Markovina S, Laszlo A, Spitz DR, Goswami PC, Yodoi J, Gius D. Thioredoxin nuclear translocation and interaction with redox factor-1 activates the activator protein-1 transcription factor in response to ionizing radiation. Cancer Res. 2000;60(23):6688–6695.

[40] Hirota K, Murata M, Sachi Y, Nakamura H, Takeuchi J, Mori K, Yodoi J. Distinct roles of thioredoxin in the cytoplasm and in the nucleus. A two-step mechanism of redox regulation of transcription factor NF-kappaB. J Biol Chem. 1999;274(39):27891–27897.

[41] Li Y, Liu X, Zhou T, Kelley MR, Edwards P, Gao H, Qiao X. Inhibition of APE1/Ref-1 redox activity rescues human retinal pigment epithelial cells from oxidative stress and reduces choroidal neovascularization. Redox Biol. 2014;2:485–494. doi:10.1016/j.redox.2014.01.023.

[42] Shan JL, He HT, Li MX, Zhu JW, Cheng Y, Hu N, Wang G, Wang D, Yang XQ et al. APE1 promotes antioxidant capacity by regulating Nrf-2 function through a redox-depend-

ent mechanism. Free Radic Biol Med. 2015;78:11–22. doi:10.1016/j.freeradbiomed.2014.10.007.

[43] Tsuji Y. JunD activates transcription of the human ferritin H gene through an antioxidant response element during oxidative stress. Oncogene. 2005;24(51):7567–7578.

[44] Iwasaki K, Hailemariam K, Tsuji Y. PIAS3 interacts with ATF1 and regulates the human ferritin H gene through an antioxidant-responsive element. J Biol Chem. 2007;282(31): 22335–22343.

[45] Kim YC, Masutani H, Yamaguchi Y, Itoh K, Yamamoto M, Yodoi J. Hemin-induced activation of the thioredoxin gene by Nrf2. A differential regulation of the antioxidant responsive element by a switch of its binding factors. J Biol Chem. 2001;276(21):18399–18406.

[46] Yu M, Li H, Liu Q, Liu F, Tang L, Li C, Yuan Y, Zhan Y, Xu W et al. Nuclear factor p65 interacts with Keap1 to repress the Nrf2-ARE pathway. Cell Signal. 2011;23(5):883–892. doi:10.1016/j.cellsig.2011.01.014.

[47] Gasdaska PY, Gasdaska JR, Cochran S, Powis G. Cloning and sequencing of a human thioredoxin reductase. FEBS Lett. 1995;373(1):5–9.

[48] Matsuzaka Y, Okamoto K, Mabuchi T, Iizuka M, Ozawa A, Oka A, Tamiya G, Kulski JK, Inoko H. Identification and characterization of novel variants of the thioredoxin reductase 3 new transcript 1 TXNRD3NT1. Mamm Genome. 2005;16(1):41–49.

[49] Reichard JF, Motz GT, Puga A. Heme oxygenase-1 induction by NRF2 requires inactivation of the transcriptional repressor BACH1. Nucleic Acids Res. 2007;35(21): 7074–7086.

[50] Jyrkkänen HK, Kuosmanen S, Heinäniemi M, Laitinen H, Kansanen E, Mella-Aho E, Leinonen H, Ylä-Herttuala S, Levonen AL. Novel insights into the regulation of antioxidant-response-element-mediated gene expression by electrophiles: induction of the transcriptional repressor BACH1 by Nrf2. Biochem J. 2011;440(2):167–174. doi: 10.1042/BJ20110526.

[51] Zolotukhin PV, Dovzhik AD, Lebedeva UA, Kuzminova ON, Mashkina EV, Aleksandrova AA, Shkurat TP. Testing the concept of the interatomic status of the NFE2L2/AP1 pathway as a systemic biomarker for examination stress. Mol Diagn Ther. 2014;18(3): 355–369. doi:10.1007/s40291-014-0088-1.

[52] Dovzhik AD, Zolotukhin PV, Mayboroda EA, Mashkina EV, Aleksandrova AA. Assessing the regulation dependencies of thioredoxin 1 transcript variants by means of interactomic dynamic-with-induction profiling approach. Living and life-remodelled systems. [Zhiviye i biokostniye sistemy]. 2014;2014(6). ID 6–8. URL: http://www.jbks.ru/archive/issue-6/article-8.

[53] DeNicola GM, Karreth FA, Humpton TJ, Gopinathan A Wei C, Frese K, Mangal D, Yu KH, Yeo CJ, Calhoun ES, et al. Oncogene-induced Nrf2 transcription promotes ROS

detoxification and tumorigenesis. Nature. 2011;475(7354):106–109. doi:10.1038/nature10189.

[54] Cho HJ, Kang JH, Kwak JY, Lee TS, Lee IS, Park NG, Nakajima H, Magae J, Chang YC. Ascofuranone suppresses PMA-mediated matrix metalloproteinase-9 gene activation through the Ras/Raf/MEK/ERK- and Ap1-dependent mechanisms. Carcinogenesis. 2007;28(5):1104–1110.

[55] Yoshioka K, Deng T, Cavigelli M, Karin M. Antitumor promotion by phenolic antioxidants: inhibition of AP-1 activity through induction of Fra expression. Proc Natl Acad Sci USA. 1995;92(11):4972–4976.

[56] Jin C, Li H, Ugai H, Murata T, Yokoyama KK. Transcriptional regulation of the c-jun gene by AP-1 repressor protein JDP2 during the differentiation of F9 cells. Nucleic Acids Res Suppl. 2002;1(2):97–98.

[57] Borde-Chiché P, Diederich M, Morceau F, Wellman M, Dicato M. Phorbol ester responsiveness of the glutathione S-transferase P1 gene promoter involves an inducible c-jun binding in human K562 leukemia cells. Leuk Res. 2001;25(3):241–247.

[58] Suzuki T, Kelly VP, Motohashi H, Nakajima O, Takahashi S, Nishimura S, Yamamoto M. Deletion of the selenocysteine tRNA gene in macrophages and liver results in compensatory gene induction of cytoprotective enzymes by Nrf2. J Biol Chem. 2008;283(4):2021–2030.

[59] Carvalho AN, Marques C, Guedes RC, Castro-Caldas M, Rodrigues E, van Horssen J, Gama MJ. S-Glutathionylation of Keap1: a new role for glutathione S-transferase pi in neuronal protection. FEBS Lett. 2016;590(10):1455–1466. doi:10.1002/1873-3468.12177.

[60] Sun Z, Wu T, Zhao F, Lau A, Birch CM, Zhang DD. KPNA6 (Importin {alpha}7)-mediated nuclear import of Keap1 represses the Nrf2-dependent antioxidant response. Mol Cell Biol. 2011;31(9):1800–1811. doi:10.1128/MCB.05036-11.

[61] Dickinson DA, Iles KE, Zhang H, Blank V, Forman HJ. Curcumin alters EpRE and AP-1 binding complexes and elevates glutamate-cysteine ligase gene expression. FASEB J. 2003;17(3):473–475.

[62] Pendyala S, Natarajan V. Redox regulation of Nox proteins. Respir Physiol Neurobiol. 2010;174(3):265–271. doi:10.1016/j.resp.2010.09.016.

[63] Pendyala S, Moitra J, Kalari S, Kleeberger SR, Zhao Y, Reddy SP, Garcia JG, Natarajan V. Nrf2 regulates hyperoxia-induced Nox4 expression in human lung endothelium: identification of functional antioxidant response elements on the Nox4 promoter. Free Radic Biol Med. 2011;50(12):1749–1759. doi:10.1016/j.freeradbiomed.2011.03.022.

[64] Shiose A, Kuroda J, Tsuruya K, Hirai M, Hirakata H, Naito S, Hattori M, Sakaki Y, Sumimoto H. A novel superoxide-producing NAD(P)H oxidase in kidney. J Biol Chem. 2001;276(2):1417–1423.

[65] Kuroda J, Nakagawa K, Yamasaki T, Nakamura K, Takeya R, Kuribayashi F, Imajoh-Ohmi S, Igarashi K, Shibata Y, Sueishi K, Sumimoto H. The superoxide-producing NAD(P)H oxidase Nox4 in the nucleus of human vascular endothelial cells. Genes Cells. 2005;10(12):1139–1151.

[66] Van Buul JD, Fernandez-Borja M, Anthony EC, Hordijk PL. Expression and localization of NOX2 and NOX4 in primary human endothelial cells. Antioxid Redox Signal. 2005;7(3–4):308–317.

[67] Bánfi B, Molnár G, Maturana A, Steger K, Hegedûs B, Demaurex N, Krause KH. A Ca(2+)-activated NADPH oxidase in testis, spleen, and lymph nodes. J Biol Chem. 2001;276(40):37594–37601.

[68] Martyn KD, Frederick LM, von Loehneysen K, Dinauer MC, Knaus UG. Functional analysis of Nox4 reveals unique characteristics compared to other NADPH oxidases. Cell Signal. 2006;18(1):69–82.

[69] Manea A, Manea SA, Gafencu AV, Raicu M, Simionescu M. AP-1-dependent transcriptional regulation of NADPH oxidase in human aortic smooth muscle cells: role of p22phox subunit. Arterioscler Thromb Vasc Biol. 2008;28(5):878–885. doi:10.1161/ATVBAHA.108.163592.

[70] Kim HJ, Jung KJ, Yu BP, Cho CG, Choi JS, Chung HY. Modulation of redox-sensitive transcription factors by calorie restriction during aging. Mech Ageing Dev. 2002;123(12):1589–1595.

[71] Jin DY, Chae HZ, Rhee SG, Jeang KT. Regulatory role for a novel human thioredoxin peroxidase in NF-kappaB activation. J Biol Chem. 1997;272(49):30952–30961.

[72] Schenk H, Klein M, Erdbrügger W, Dröge W, Schulze-Osthoff K. Distinct effects of thioredoxin and antioxidants on the activation of transcription factors NF-kappaB and AP-1. Proc Natl Acad Sci U S A. 1994;91(5):1672–1676.

[73] Nordberg J, Arnér ES. Reactive oxygen species, antioxidants, and the mammalian thioredoxin system. Free Radic Biol Med. 2001;31(11):1287–1312.

[74] Wu JX, Carpenter PM, Gresens C, Keh R, Niman H, Morris JW, Mercola D. The proto-oncogene c-fos is over-expressed in the majority of human osteosarcomas. Oncogene. 1990;5(7):989–1000.

[75] Bohmann D, Bos TJ, Admon A, Nishimura T, Vogt PK, Tjian R. Human proto-oncogene c-jun encodes a DNA binding protein with structural and functional properties of transcription factor AP-1. Science. 987;238(4832):1386–1392.

[76] Jeong HS, Ryoo IG, Kwak MK. Regulation of the expression of renal drug transporters in KEAP1-knockdown human tubular cells. Toxicol In Vitro. 2015;29(5):884–892. doi: 10.1016/j.tiv.2015.03.013.

[77] Rao S, Tortola L, Perlot T, Wirnsberger G, Novatchkova M, Nitsch R, Sykacek P, Frank L, Schramek D, Komnenovic V et al. A dual role for autophagy in a murine model of lung cancer. Nat Commun. 2014;5:3056. doi:10.1038/ncomms4056.

[78] Sankaranarayanan K, Jaiswal AK. Nrf3 negatively regulates antioxidant-response element-mediated expression and antioxidant induction of NAD(P)H:quinone oxidoreductase1 gene. J Biol Chem. 2004;279(49):50810–50817.

[79] Chevillard G, Blank V. NFE2L3 (NRF3): the Cinderella of the Cap'n'Collar transcription factors. Cell Mol Life Sci. 2011;68(20):3337–3348. doi:10.1007/s00018-011-0747-x.

[80] Derjuga A, Gourley TS, Holm TM, Heng HH, Shivdasani RA, Ahmed R, Andrews NC, Blank V. Complexity of CNC transcription factors as revealed by gene targeting of the Nrf3 locus. Mol Cell Biol. 2004;24(8):3286–3294.

[81] Zhang Y, Kobayashi A, Yamamoto M, Hayes JD. The Nrf3 transcription factor is a membrane-bound glycoprotein targeted to the endoplasmic reticulum through its N-terminal homology box 1 sequence. J Biol Chem. 2009;284(5):3195–3210. doi:10.1074/jbc.M805337200.

[82] Pepe AE, Xiao Q, Zampetaki A, Zhang Z, Kobayashi A, Hu Y, Xu Q. Crucial role of nrf3 in smooth muscle cell differentiation from stem cells. Circ Res. 2010;106(5):870–879. doi: 10.1161/CIRCRESAHA.109.211417.

[83] Burton GJ, Jauniaux E. Oxidative stress. Best Pract Res Clin Obstet Gynaecol. 2011;25(3): 287–299. doi:10.1016/j.bpobgyn.2010.10.016.

[84] Chepelev NL, Bennitz JD, Huang T, McBride S, Willmore WG. The Nrf1 CNC-bZIP protein is regulated by the proteasome and activated by hypoxia. PLoS One. 2011;6(12):e29167. doi:10.1371/journal.pone.0029167.

[85] Lee OH, Jain AK, Papusha V, Jaiswal AK. An auto-regulatory loop between stress sensors INrf2 and Nrf2 controls their cellular abundance. J Biol Chem. 2007;282(50): 36412–36420.

[86] Kaspar JW, Niture SK, Jaiswal AK. Nrf2:INrf2 (Keap1) signaling in oxidative stress. Free Radic Biol Med. 2009;47(9):1304–1309. doi:10.1016/j.freeradbiomed.2009.07.035.

[87] Piantadosi CA, Carraway MS, Babiker A, Suliman HB. Heme oxygenase-1 regulates cardiac mitochondrial biogenesis via Nrf2-mediated transcriptional control of nuclear respiratory factor-1. Circ Res. 2008;103(11):1232–1240. doi:10.1161/01.RES.0000338597.71702.ad.

[88] Jain AK, Jaiswal AK. GSK-3beta acts upstream of Fyn kinase in regulation of nuclear export and degradation of NF-E2 related factor 2. J Biol Chem. 2007;282(22):16502–16510.

[89] Niture SK, Kaspar JW, Shen J, Jaiswal AK. Nrf2 signaling and cell survival. Toxicol Appl Pharmacol. 2010;244(1):37–42. doi:10.1016/j.taap.2009.06.009.

[90] Liu GH, Qu J, Shen X. NF-kappaB/p65 antagonizes Nrf2-ARE pathway by depriving CBP from Nrf2 and facilitating recruitment of HDAC3 to MafK. Biochim Biophys Acta. 2008;1783(5):713–727. doi:10.1016/j.bbamcr.2008.01.002.

[91] Lee DF, Kuo HP, Liu M, Chou CK, Xia W, Du Y, Shen J, Chen CT, Huo L, Hsu MC et al. KEAP1 E3 ligase-mediated downregulation of NF-kappaB signaling by targeting IKKbeta. Mol Cell. 2009;36(1):131–140. doi:10.1016/j.molcel.2009.07.025.

[92] Hoshino H, Kobayashi A, Yoshida M, Kudo N, Oyake T, Motohashi H, Hayashi N, Yamamoto M, Igarashi K. Oxidative stress abolishes leptomycin B-sensitive nuclear export of transcription repressor Bach2 that counteracts activation of Maf recognition element. J Biol Chem. 2000;275(20):15370–15376.

[93] Ishikawa M, Numazawa S, Yoshida T. Redox regulation of the transcriptional repressor Bach1. Free Radic Biol Med. 2005;38(10):1344–52.

[94] Klaassen CD, Reisman SA. Nrf2 the rescue: effects of the antioxidative/electrophilic response on the liver. Toxicol Appl Pharmacol. 2010;244(1):57–65. doi:10.1016/j.taap.2010.01.013.

[95] Zhong JL, Raval C, Edwards GP, Tyrrell RM. A role for Bach1 and HO-2 in suppression of basal and UVA-induced HO-1 expression in human keratinocytes. Free Radic Biol Med. 2010;48(2):196–206. doi:10.1016/j.freeradbiomed.2009.10.037.

[96] Dhakshinamoorthy S, Jain AK, Bloom DA, Jaiswal AK. Bach1 competes with Nrf2 leading to negative regulation of the antioxidant response element (ARE)-mediated NAD(P)H:quinone oxidoreductase 1 gene expression and induction in response to antioxidants. J Biol Chem. 2005;280(17):16891–16900.

[97] Kaspar JW, Jaiswal AK. Antioxidant-induced phosphorylation of tyrosine 486 leads to rapid nuclear export of Bach1 that allows Nrf2 to bind to the antioxidant response element and activate defensive gene expression. J Biol Chem. 2010;285(1):153–162. doi:10.1074/jbc.M109.040022.

[98] Suzuki H, Tashiro S, Hira S, Sun J, Yamazaki C, Zenke Y, Ikeda-Saito M, Yoshida M, Igarashi K. Heme regulates gene expression by triggering Crm1-dependent nuclear export of Bach1. EMBO J. 2004;23(13):2544–2553.

[99] Vriend J, Reiter RJ. The Keap1-Nrf2-antioxidant response element pathway: a review of its regulation by melatonin and the proteasome. Mol Cell Endocrinol. 2015;401:213–220. doi:10.1016/j.mce.2014.12.013.

[100] Thomsen MK, Bakiri L, Hasenfuss SC, Hamacher R, Martinez L, Wagner EF. JUNB/AP-1 controls IFN-γ during inflammatory liver disease. J Clin Invest. 2013;123(12): 5258–5268.doi:10.1172/JCI70405.

[101] Realini C, Dubiel W, Pratt G, Ferrell K, Rechsteiner M. Molecular cloning and expression of a gamma-interferon-inducible activator of the multicatalytic protease. J Biol Chem. 1994;269(32):20727–20732.

[102] Ahn JY, Tanahashi N, Akiyama K, Hisamatsu H,Noda C, Tanaka K, Chung CH, Shibmara N, Willy PJ, Mott JD, et al. Primary structures of two homologous subunits of PA28, a gamma-interferon-inducible protein activator of the 20S proteasome. FEBS Lett. 1995;366(1):37–42.

[103] Cheng C, Li W, Zhang Z, Yoshimura S, Hao Q, Zhang C, Wang Z. MicroRNA-144 is regulated by activator protein-1 (AP-1) and decreases expression of Alzheimer disease-related a disintegrin and metalloprotease 10 (ADAM10). J Biol Chem. 2013;288(19): 13748–13761. doi:10.1074/jbc.M112.381392.

[104] Sangokoya C, Telen MJ, Chi JT. microRNA miR-144 modulates oxidative stress tolerance and associates with anemia severity in sickle cell disease. Blood. 2010;116(20): 4338–4348. doi:10.1182/blood-2009-04-214817.

[105] Papp D, Lenti K, Módos D, Fazekas D, Dúl Z, Türei D, Földvári-Nagy L, Nussinov R, Csermely P, Korcsmáros T. The NRF2-related interactome and regulome contain multifunctional proteins and fine-tuned autoregulatory loops. FEBS Lett. 2012;586(13): 1795–1802. doi:10.1016/j.febslet.2012.05.016.

[106] Zolotukhin PV, Belanova AA, Prazdnova EV, Mazanko MS, Batiushin MM, Chmyhalo VK, Chistyakov VA. Mitochondria as a signaling Hub and target for phenoptosis shutdown. Biochemistry (Mosc). 2016;81(4):329–337. doi:10.1134/S0006297916040039.

[107] Miyamoto N, Izumi H, Miyamoto R, Kondo H, Tawara A, Sasaguri Y, Kohno K. Quercetin induces the expression of peroxiredoxins 3 and 5 via the Nrf2/NRF1 transcription pathway. Invest Ophthalmol Vis Sci. 2011;52(2):1055–1063. doi:10.1167/iovs.10-5777.

[108] Klinge CM. Estrogenic control of mitochondrial function and biogenesis. J Cell Biochem. 2008;105(6):1342–1351. doi:10.1002/jcb.21936.

[109] Webb P, Nguyen P, Valentine C, Lopez GN, Kwok GR, McInerney E, Katzenellenbogen BS, Enmark E, Gustafsson JA, Nilsson S et al. The estrogen receptor enhances AP-1 activity by two distinct mechanisms with different requirements for receptor transactivation functions. Mol Endocrinol. 1999;13(10):1672–1685.

[110] Ross D, Kepa JK, Winski SL, Beall HD, Anwar A, Siegel D. NAD(P)H:quinone oxidoreductase 1 (NQO1): chemoprotection, bioactivation, gene regulation and genetic polymorphisms. Chem Biol Interact. 2000;129(1–2):77–97.

[111] Dahlman-Wright K, Qiao Y, Jonsson P, Gustafsson JÅ, Williams C, Zhao C. Interplay between AP-1 and estrogen receptor α in regulating gene expression and proliferation

networks in breast cancer cells. Carcinogenesis. 2012;33(9):1684–1691. doi:10.1093/carcin/bgs223

[112] Lo SC, Hannink M. PGAM5 tethers a ternary complex containing Keap1 and Nrf2 to mitochondria. Exp Cell Res. 2008;314(8):1789–1803. doi:10.1016/j.yexcr.2008.02.014.

[113] Lu W, Sun J, Yoon JS, Zhang Y, Zheng L, Murphy E, Mattson MP, Lenardo MJ. Mitochondrial Protein PGAM5 Regulates Mitophagic Protection against Cell Necroptosis. PLoS One. 2016;11(1):e0147792. doi:10.1371/journal.pone.0147792. eCollection 2016.

[114] Geisler S, Holmström KM, Skujat D, Fiesel FC, Rothfuss OC, Kahle PJ, Springer W. PINK1/Parkin-mediated mitophagy is dependent on VDAC1 and p62/SQSTM1. Nat Cell Biol. 12(2):119–131. doi:10.1038/ncb2012.

[115] Fratta Pasini A, Albiero A, Stranieri C, Cominacini M, Pasini A, Mozzini C, Vallerio P, Cominacini L, Garbin U. Serum oxidative stress-induced repression of Nrf2 and GSH depletion: a mechanism potentially involved in endothelial dysfunction of young smokers. PLoS One. 2012;7(1):e30291. doi:10.1371/journal.pone.0030291.

[116] Sirianni R, Nogueira E, Bassett MH, Carr BR, Suzuki T, Pezzi V, Andò S, Rainey WE. The AP-1 family member FOS blocks transcriptional activity of the nuclear receptor steroidogenic factor 1. J Cell Sci. 2010;123(Pt 22):3956–3965. doi:10.1242/jcs.055806.

[117] Venugopal R, Jaiswal AK. Nrf1 and Nrf2 positively and c-Fos and Fra1 negatively regulate the human antioxidant response element-mediated expression of NAD(P)H:quinone oxidoreductase1 gene. Proc Natl Acad Sci U S A. 1996;93(25):14960–14965.

[118] Ohtsubo T, Kamada S, Mikami T, Murakami H, Tsujimoto Y. Identification of NRF2, a member of the NF-E2 family of transcription factors, as a substrate for caspase-3(-like) proteases. Cell Death Differ. 1999;6(9):865–872.

[119] Shukla SJ, Huang R, Simmons SO, Tice RR, Witt KL, Vanleer D, Ramabhadran R, Austin CP, Xia M. Profiling environmental chemicals for activity in the antioxidant response element signaling pathway using a high throughput screening approach. Environ Health Perspect. 2012;120(8):1150–1156. doi:10.1289/ehp.1104709.

[120] Belanova AA, Smirnov DS, Makarenko MS, Belousova MM, Aleksandrova AA, Soldatov AV, Zolotukhin PV. Individual expression features of GPX2, NQO1 and SQSTM1 transcript variants induced by hydrogen peroxide treatment in HeLa cells. Genetics and molecular biology. 2015. Publication in progress, major revision submitted.

[121] Crunkhorn S. Deal watch: Abbott boosts investment in NRF2 activators for reducing oxidative stress. Nat Rev Drug Discov. 2012;11(2):96. doi:10.1038/nrd36

Section 2

Participation of Nrf2 in Diseases

Chapter 5

Nrf2 Contributes to the Poor Prognosis and Chemoresistance

Jhih-Pu Syu, Jen-Tsan Chi and Hsiu-Ni Kung

Additional information is available at the end of the chapter

http://dx.doi.org/10.5772/65105

Abstract

With the increasing incidence of human cancer and continued difficulty in treating metastatic tumors, there is an urgent need to identify biomarkers for tumors with poor outcome and novel therapeutic targets. Many therapeutic targets have been found in recent years. One promising biomarker and therapeutic target that is valuable for human tumor is nuclear factor erythroid 2-related factor 2 (NFE2L2, Nrf2). In this chapter, we will discuss the regulatory mechanisms and conflicting roles of Nrf2 during different stages of tumor development as well as its involvement in the drug resistance and hypoxia-induced chemoresistance. We will also discuss various positive and negative modulators of Nrf2 as reference to their potential utility as study tools and leads for further clinical development.

Keywords: Nrf2, oxidative stress, ROS, tumor, prognosis, chemoresistance

1. Introduction

Nrf2 is the main regulator for the expression of antioxidant enzymes and the detoxification proteins. With these abilities of Nrf2, Nrf2 activation confers cells with more anti-stress capacity, thus resulting in more malignancy and chemoresistance of tumor cells. Therefore, targeting Nrf2 in tumor may offer therapeutic benefit by undermining its advantage on the proliferation, migration, metastasis, and drug resistance of tumor cells. Collectively, Nrf2 has the potential to serve as a good biomarker and therapeutic target to overcome the poor prognosis and chemoresistance associated with tumor or tumor hypoxia [1, 2].

2. Nrf2 is the key regulator of antioxidant and detoxification abilities

Organisms survive based on the normal and steady function of cell metabolism to maintain cellular homeostasis. However, the process of metabolism produces many metabolic wastes. If these wastes are not properly removed, they become harmful. One of these metabolic wastes is the reactive oxygen species (ROS). While ROS can serve the signaling function, uncontrolled ROS may lead to cellular damage and death. Many key enzymes involved in the removal of excess ROS are tightly regulated by a transcription factor, Nrf2.

Nrf2 activation is strictly regulated by an ubiquitin-proteosome system (UPS) [3]. Nrf2 is negatively regulated by Keap1 (Kelch-like ECH-associated protein 1), the most important molecular switch controlling the activation and inactivation of the Nrf2 pathway. Keap1 is an adaptor for Nrf2 as the Nrf2-Keap1 complex becomes a substrate for the Cul3-dependent E3 ubiquitin ligase for proteasomal degradation. Keap1 is a cysteine-rich protein, which can be modified by many oxidants and electrophiles. Upon exposure to oxidative stress, these cysteine residues may be altered by stresses to induce conformational changes that inhibit Keap1-dependent ubiquitin ligase activity to allow Nrf2 accumulation [4, 5].

Since Nrf2 is a transcription factor, the accumulated Nrf2 proteins translocate into the nucleus to dimerize with members of the small Maf family, and bind to the antioxidant response element (ARE or electrophile response element 5′-RTGABNNNGCR-3′) in the promoter regions of cell defense genes [6]. These Nrf2-regulated proteins include Phase II detoxification enzymes and some stress response proteins as listed below:

1. Antioxidant proteins: proteins that control the antioxidant specializing in neutralizing the reactive species and protecting organisms from oxidative damage, such as NAD(P)H: quinone oxidoreductase 1 (NQO1), epoxide hydrolase, heme oxygenase 1 (HO-1), glutathione *S*-transferase (GST), and glutathione peroxidase (GPx).

2. Glutathione producing enzymes: proteins that regulate the synthesis and metabolism of glutathione, such as glutamate-cysteine ligase (GCL), which consists of two subunits, a light regulatory subunit; glutamate-cysteine ligase modifier subunit (GCLM), and a heavy catalytic enzyme; GCL catalytic subunit (GCLC).

3. Drug-metabolizing enzymes: enzymes that regulate the metabolism of drugs, including UDP-glucuronosyl-transferase 1A1 (UGT1A), carbonyl reductase 1 (CBR1), aldo-keto reductases (AKR), and cytochrome P450 (CYPs).

4. Xenobiotic transporters (ATP-binding cassette (ABC) transporter): proteins that belong to ATP phosphohydrolase, some of them are involved in the exclusion of drugs, xenobiotics and their metabolites [7], which are named multidrug resistance proteins, such as multidrug resistance protein 1 (MRP1).

5. Numerous other stress response proteins, such as thioredoxin, ferritin subunits, and copper/zinc superoxide dismutase [8] (see **Figure 1**).

http://dx.doi.org/10.5772/65105

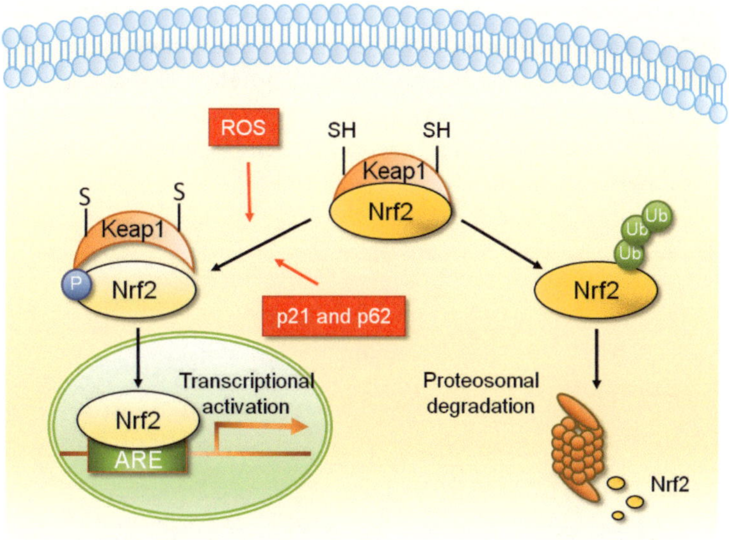

Figure 1. Regulation of Nrf2. Keap1 binds to Nrf2 and sends it to proteasomal degradation. ROS can lead to the conformational change of Keap1 and Nrf2 release. Free Nrf2 then enters the nucleus and initiate the following transcriptional process by binding to the ARE domains, P21 and P62.

3. The conflicting roles of Nrf2

Since Nrf2 is the key regulator of antioxidant capacity and detoxified proteins, the activation of Nrf2 is expected to protect cells from stresses, such as reactive oxygen species (ROS). Therefore, the Nrf2 pathway is so called oxidative stress response pathway or cellular defense pathway. Once cells or organisms are exposed to ROS induced by physical forces or chemicals, Nrf2 is activated to increase the anti-stress capacity and cope with the ROS. Nrf2 activation can stabilize the intracellular oxidant level and maintain the redox state within cells to avoid DNA damages, genomic instability, and potentially serious sabotages caused by ROS [9]. Although DNA repair mechanism can reduce slight DNA damage, higher presence of oxidizing base lesions in DNA leads to DNA mutation, which may cause aging [10], cell damage, cell death, carcinogenesis, and even cancer. Therefore, various Nrf2 activators are being pursued in chemopreventive strategies [11] to reduce tumor development. In addition, Nrf2 activators have been used to treat various human diseases, including diabetic nephropathy [12, 13] and sickle cell disease [14].

On the other hand, excess ROS may lead to numerous diseases, such as inflammation, obesity, and other metabolic diseases. For example, too much oxidative stress affects the differentiation of adipocytes and impairs the normal function of white adipose tissue [15], leading to inflammation and adipokine secretion that affect the whole organism [16, 17]. The activation of Nrf2 defense pathway can protect organisms from many metabolic diseases.

Therefore, Nrf2 can be the double edged sword in the organism. In normal cells, Nrf2 activation keeps the redox homeostasis and prevents cancer development. However, once cancer cells have established, Nrf2 activation may drive oncogenesis and confer chemoresistance. In many cancers, constitutive Nrf2 activation is an oncogenic mutation [18] and a biomarker for poor prognosis [19, 20] (**Figure 2**). In the TCGA data, mutations in the Nrf2 pathways constitute one of the major oncogenic pathways of lung cancers [21, 22]. The angel and devil roles of Nrf2 are discussed in the following sections.

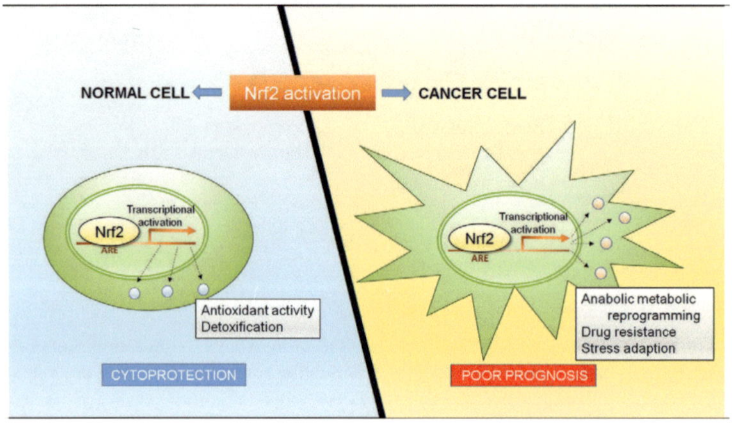

Figure 2. Nrf2 produces phase II enzymes that provide cell defense system, such as the antioxidant and detoxification, in normal cells. Once tumor is formed and cancer cells get the cytoprotective abilities of Nrf2, which triggers anabolic metabolic reprogramming, drug resistance, and stress adaption, Nrf2 leads to poor prognosis in patients.

4. The good side of Nrf2

Nrf2 activation in normal cells makes cells stronger against environmental stresses and prevents carcinogenesis. Nrf2 is able to augment a wide range of cell defense processes, thereby enhancing the overall capacity of cells to detoxify potentially harmful entities. As such, the Keap1-Nrf2 pathway is generally considered as the major cellular defense pathway that offers survival advantages.

Keap1-Nrf2 is the key cellular defense mechanism to combat oxidative stress. The activated Nrf2 protect organisms from these diseases by diminishing the ROS. Nrf2 is such a natural cytoprotective response against oxidative stress-induced inflammation. Nrf2-null mice tend to spontaneously develop various inflammatory disorders, including glomerulonephritis [23], immune-mediated hemolytic anemia [24], and multiorgan autoimmune inflammation [25]. Also, activated Nrf2 protects many body systems, including airway, liver, gastrointestinal tract and kidney, where these systems are attacked by toxic agents very often [26]. For example, Nrf2 activation via sulforaphane (Nrf2 inducer) protects kidney from chronic renal disease [27] by increasing the GCLC and glutathione level. Activation of Nrf2 also alleviates the TGF-β-induced, increased α-SMA and repressed E-cadherin [28], which are the markers for epithelial-mesenchymal transition (EMT), through the SMUR1-SMAD7 signaling. Another Nrf2 inducer, AST120, can restore the HO-1 and NQO1 levels and decrease the production of ROS stimulated by indoxyl sulfate-induced chronic renal disease [29] Nrf2 not only protects the kidney but also protects the lungs. Nrf2 protects lungs from chronic pulmonary injury [30], fibrosis [31], and acute lung injury [32, 33]. Therefore, Nrf2 is named as the "multiorgan protector" [34].

With cytoprotective functions and the cellular defense mechanism against exogenous and endogenous insults, Nrf2 is considered as a tumor suppressor. In one hand, *in vivo* tumor development data using Nrf2-knockout mice has highlighted the tumor suppression ability of Nrf2. With treatment of chemical and physical stimuli, Nrf2-null mice are more prone to develop cancer [35]. On the other hand, Nrf2 activation can remove damaged proteins, promoting the overall survival of the cell and detoxify the cellular environment to maintain the homeostasis in the organism [36]. The abilities of Nrf2 to combat oxidative stress and inflammation, which are conducive to initiate oncogenesis, attain a result of tumor suppression [36].

5. The dark side of Nrf2

Gain of Nrf2 in cancer cells: the Nrf2 pathway is a powerful sensor for cellular redox state and is activated directly by oxidative stress and/or indirectly by stress response protein kinases. Although Nrf2 is beneficial to normal cells to fight against stresses, once tumor cells get the antioxidant and detoxificative abilities of Nrf2, things go in another direction. For example, the constitutive Nrf2 activation has redirected tumor metabolism to support the biosynthetic needs of tumor proliferation [37, 38]. In addition, Nrf2 makes cancer cells stronger against chemotherapy and leads cells to become more malignant. In this case, Nrf2 serves as a target for chemotherapy. Recent researches have highlighted that persistent accumulation of Nrf2 in cancer cells is harmful, since it can promote the survival and proliferation of cells that have acquired cancer-promoting mutations, and Nrf2 is also observed beneficial for tumorigenesis [11, 39–42]. Nrf2 orchestrates the expression of various genes that help cancer cells to resist chemotherapeutic treatment, including antioxidants (NQO1, NQO2, HO-1, and GCLC), antiapoptotic (Bcl-2), drug-metabolizing enzymes (G6PD, TKT, and PPARγ), and drug efflux transporters (ABCG2, MRP3, and MRP4) genes [43].

The activation of Nrf2-ARE pathway protects cancer cells from oxidative toxicity and H_2O_2-induced apoptosis [44, 45]. The effects of Nrf2 on tumors or cancer cells are listed below:

A. Proliferation, tumorigenesis and poor patient survival: Nrf2 contributes to the tumorigenesis, cancer proliferation in bench and poor patient survival in clinic in various tumors, including hepatocarcinoma (HCC) [46], breast tumor [47], nonsmall cell lung cancer (NSCLC) [48], glioma [49, 50], pancreatic adenocarcinoma [51], and gastric cancer [52]. Nrf2 is also found to involve in the maintenance of quiescence, survival, and stress resistance of cancer stem cells (CSCs), thus dedicating to tumor progression and recurrence [53].

B. Chemotherapeutic resistance: Nrf2 exerts the detoxification and drug export through activating multidrug resistance proteins and drug transporters. This action protects cancer cells from the damage of chemotherapy, such as 5-fluorouracil (5-FU) in gastric cancer [54] and in gallbladder cancer [55], and cisplatin (CDDP) and camptothecin in pancreatic cancer [56].

C. Epithelial-mesenchymal transition (EMT), tumor metastasis, and malignancy: cancer cells respond to some anti-diabetic drugs, which have antioxidant properties and inhibit the Kea1-dependent obstruction of Nrf2, with Nrf2 activation and result in increased migration and metastasis, such as hypoglycemic dipeptidyl peptidase-4 inhibitors (DPP-4i), saxagliptin and sitagliptin [57].

D. Hypoxia-induced drug resistance: Nrf2 contributes to the chemotherapeutic drug resistance induced by hypoxia in breast cancers [2]. Hypoxia is a natural status in the tumor center where the cancer cells outgrow the perfusion from local blood vessels for getting enough oxygen and nutrients. Hypoxia triggers the ROS unbalance and stimulates the activation of Nrf2 [2]. Following the Nrf2 nuclear translocation and ARE binding, antioxidant enzymes are produced to maintain the stability of intracellular redox state. Blocking ROS unbalance by ROS scavenger inhibits the Nrf2 activation and the following drug resistance. Inhibition of Nrf2 or the production of related enzymes with siRNA or specific inhibitor blocks the chemoresistance under hypoxia (**Figure 3**).

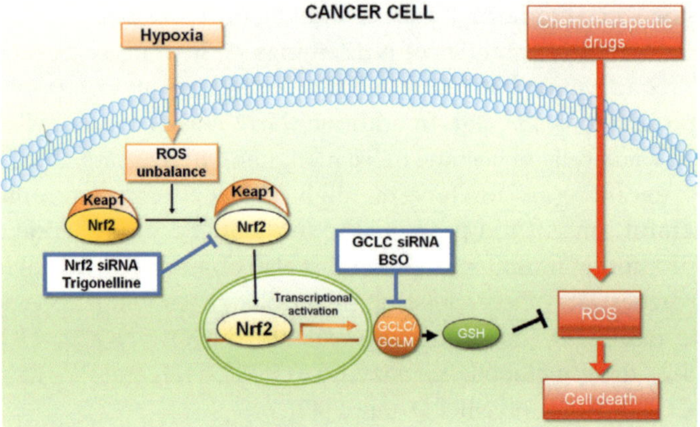

Figure 3. Nrf2, activated by hypoxia-induced ROS unbalance, starts the production of protective enzymes and contributes to the hypoxia-related chemoresistance. Specific inhibitors or siRNA targeting Nrf2 or downstream enzymes increases the cytotoxic effects of chemotherapeutic drugs.

6. Nrf2 activation in tumors

Nrf2 gets activated in malignant tumors, such as carcinomas of skin, lung, oesophagus, and larynx [58]. Many factors control the activation of Nrf2 in tumors, some are listed below.

http://dx.doi.org/10.5772/65105

a. The somatic mutation of Keap1 or Nrf2: the mutations in Keap1 or Nrf2 hurt the interaction of Keap1 and Nrf2, leading to a higher free Nrf2 level and activity [59].

b. The decreased level of Keap1: besides somatic mutations of Keap1, epigenetic changes, such as hypermethylation on the promoter region of Keap1, decrease the expression level of Keap1 [60]. In addition, several studies have shown that miRNAs, including mir-200A [61, 62], miR-141 [63] and mir-28 [64], also regulates the expression level of Keap1 mRNA. These events lead to the decreased level of Keap1 and the nuclear accumulation of Nrf2.

c. The increased level of Nrf2: Nrf2 can be activated by increase of some oncogenes, such as Kras^{G12D} [42], or disruption of tumor suppressors, such as PTEN [37] in tumors, that lead to better cell survival and higher drug resistance.

d. Nrf2 polymorphism: in addition to varying Nrf2 expression, Nrf2 polymorphism also affects the Nrf2 activity. The Nrf2 polymorphism contributes to poor prognosis in cancers, including cholangiocarcinoma [65], lung cancer [66], and breast cancer [67]. It also contributed to diseases, such as increasing the risk of acute lung injury [68, 69] as well as blood pressure and cardiovascular mortality in patients with hemodialysis [70].

7. Regulators of Nrf2

Many molecules and chemicals are thought to regulate the Nrf2 activation; some of them are described as following (see **Figure 4**):

A. Negative regulators:

♦ **Endogenous**

a. Keap1: Keap1 is a natural intracellular molecule that negatively regulates the activation of Nrf2. Keap1 binds to Nrf2 in the cytoplasm, sends Nrf2 to proteosome digestion, and keeps a low Nrf2 level in cells.

b. Ubiquitin-specific processing protease 15 (USP15): USP15 deubiquitinates Keap1, stabilizes the Keap1-Cul3-E3 ligase complex, and enhances the E3 ligase activity, which leads to the binding between Keap1 and Nrf2 and the degradation of Nrf2 [71].

♦ **Exogenous**

c. Trigonelline: trigonelline is a coffee alkaloids that reduce nuclear accumulation of the Nrf2 protein, block expression of proteasomal genes (for example, s5a/psmd4 and α5/psma5), and reduce proteasome activity regulated by Nrf2 [72].

d. Vitamin C (ascorbic acid) [73, 74] and Vitamin E [75]: these two vitamins are water soluble vitamins with a high capacity to capture ROS. Since Nrf2 is activated via ROS imbalance, elimination of ROS by vitamins keeps low intracellular ROS and ends in low Nrf2 activity.

e. ROS scavenger: ROS scavenger is a common name for molecules that can balance the intracellular ROS, including dithiothreitol (DTT) [76], N-acetylcysteine (NAC) [77], catalase [78], 6-hydroxy-2,5,7,8-tetramethylchromane-2-carboxylic acid (Trolox) [75]. They are used as standards to check the antioxidant capacity of other molecules. Nrf2 activity remains low where these ROS scavengers keep the intracellular ROS level steady.

f. Brusatol: Brusatol is a quassinoid that provokes a rapid and transient inhibition of Nrf2 signaling. It increases the intracellular oxidative stress via inhibition of Nrf2 [79–81].

B. Activators:

♦ **Stresses**

a. ROS: ROS imbalance is the key regulator of Nrf2. Excess ROS induces the activation and nuclear translocation of Nrf2 to keep intracellular ROS balance through upregulating the level and activity of antioxidant enzymes.

b. Hypoxia: hypoxia or lack of oxygen is a prominent tumor environmental stress, and reported to induce ROS unbalance that is subsequently leading to Nrf2 activation [2, 82, 83].

♦ **Disruptor proteins and transcription factors**

c. P21 and p62: in addition to the conformational change of Keap1 to loss the bonding affinity to Nrf2, some proteins, such as p21 and p62, can directly bond to Keap1 or Nrf2, disrupting the interaction between Nrf2 and Keap1, thus ending in the nuclear accumulation and activation of Nrf2 [84–86].

d. AhR: Aryl hydrocarbon receptor (AhR), which is a ligand-dependent transcription factor by forming a heterodimer with the aryl hydrocarbon nuclear translocator (Arnt) as a nuclear partner protein. The heterodimeric protein complex regulates expression of Nrf2, and promotes the expression of phase I enzymes, phase II enzymes, and multidrug resistance-associated proteins [87–90]. Nrf2 can also regulate the activation of AhR and subsequently modulates downstream AhR signaling cascades, including increasing the expression of xenobiotic metabolism genes and inhibit the adipogenesis in mouse embryonic fibroblasts (MEFs) [91].

e. Ebselen: ebselen is a glutathione peroxidase-1 mimetic and a seleno-organic antioxidant. It attenuated cisplatin-induced oxidative stress generation through Nrf2 pathway [92].

♦ **Natural products or extracts**

f. Sulforaphane (SFN): sulforaphane, which is found in cruciferous vegetables, belongs to the isothiocyanate family (such as broccoli) and is widely used as an antioxidant supplement and applied in cancer chemoprevention [93]. SFN reacts with Keap1 and block the binding of Keap1 and Nrf2, thus activates Nrf2 and the antioxidant function [94].

g. Curcumin: curcumin, which is a polyphenolic natural extract of turmeric [95], is reported to exhibit anti-inflammation and antitumorigenic activity and chemoprevention effect [96, 97]. To activate those protective proteins, curcumin increase the antioxidant genes through regulating the binding of Nrf2 and ARE [98, 99]. Thus, curcumin becomes one of the Nrf2 activator.

h. Resveratrol: in addition to curcumin, resveratrol is another plant extract that regulates antioxidant ability. Resveratrol, the extract from grapes, berries and peanuts, exerts antioxidant, anti-inflammation, and anti-aging effects in experimental animals. Resveratrol is also reported to upregulate Nrf2 activity in cells and organisms to elevate the protection effects toward environmental stresses [100].

i. Coffee: coffee is one of the most widely consumed beverages in the world. Coffee is noted for its antioxidant ability which protects against chronic liver disease, diabetes, and hepatocarcinoma development with the right amount. The antioxidant ability of coffee is through the activation of Nrf2 and AhR to protect organs from oxidative stress, at least in liver and stomach [101].

j. Caffeic acid phenethyl ester (CAPE): CAPE, a major component extracted from the bee product propolis in honeybee hives, is known to have antimitogenic, anticarcinogenic, anti-inflammatory activities. It activates the Nrf2 pathway to inhibit oxidative stress and inflammation [102].

k. Cinnamic aldehyde: cinnamic aldehyde, which is found in cinnamon bark, enhances Nrf2 nuclear translocation and activates Nrf2 -dependent antioxidant response to overcome stresses [103, 104].

l. Flavonoid (Chrysin, Apigenin, Luteolin): these three flavonoids can reduce the ROS level through activating Nrf2 and producing the downstream phase II enzymes [105].

♦ **Synthesized compounds**

m. Oltipraz: oltipraz is an organosulfur compound belonging to the dithiolethione class. It is also a bifunctional inducer activating both phase I and phase II drug-metabolizing enzymes via the xenobiotic responsive element [106]. It has been used as an Nrf2 activator in recent studies [107, 108].

n. Tertiary butylhydroquinone (tBHQ): tBHQ, the major metabolite of butylated hydroxyanisole, stabilizes Nrf2 and induces Nrf2 activation through mitochondrial oxidative stress induction [109, 110].

o. Other food and clinical drugs: many other foods or clinical drugs can also affect the expression and activity of Nrf2 in recent studies.

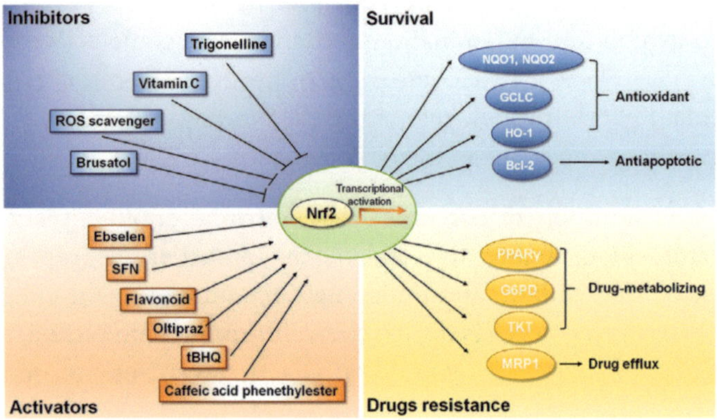

Figure 4. Positive and negative regulators of Nrf2, and the functions of enzymes and proteins produced by Nrf2 activation.

8. Conclusion

Nrf2 is powerful in the cell defense system toward oxidative stress caused by various physiological and chemical stresses. Nrf2 activation benefits the survival of not only normal cells but also cancer cells. With the anti-oxidation and detoxification abilities of Nrf2, the proliferation, tumorigenicity, migration, and metastasis of cancer cells are higher. The detoxification and drug export mechanism also give cancer cells the ability to fight against chemotherapeutic drugs. With all the characteristics of Nrf2, it is a good marker for both poor prognosis and drug resistance in tumors, both in the regular normoxic environment or under hypoxic environment [111].

Author details

Jhih-Pu Syu[1], Jen-Tsan Chi[2,3] and Hsiu-Ni Kung[1*]

*Address all correspondence to: kunghsiuni@gmail.com

1 Department of Anatomy and Cell Biology, College of Medicine, National Taiwan University, Taipei, Taiwan

2 Center for Genomic and Computational Biology, Duke University, Durham, NC, USA

3 Department of Molecular Genetics and Microbiology, Duke University, Durham, NC, USA

References

[1] Zhang, M., Zhang, C., Zhang, L., Yang, Q., Zhou, S., Wen, Q., and Wang, J. (2015) Nrf2 is a potential prognostic marker and promotes proliferation and invasion in human hepatocellular carcinoma. *BMC Cancer* 15, 531.

[2] Syu, J. P., Chi, J. T., and Kung, H. N. (2016) Nrf2 is the key to chemotherapy resistance in MCF7 breast cancer cells under hypoxia. *Oncotarget* 7, 14659–14672.

[3] Chapple, S. J., Siow, R. C., and Mann, G. E. (2012) Crosstalk between Nrf2 and the proteasome: therapeutic potential of Nrf2 inducers in vascular disease and aging. *Int J Biochem Cell Biol* 44, 1315–1320.

[4] Canning, P., Sorrell, F. J., and Bullock, A. N. (2015) Structural basis of Keap1 interactions with Nrf2. *Free Radic Biol Med* 88, 101–107.

[5] Vriend, J., and Reiter, R. J. (2015) The Keap1-Nrf2-antioxidant response element pathway: a review of its regulation by melatonin and the proteasome. *Mol Cell Endocrinol* 401, 213–220.

[6] Lacher, S. E., Lee, J. S., Wang, X., Campbell, M. R., Bell, D. A., and Slattery, M. (2015) Beyond antioxidant genes in the ancient Nrf2 regulatory network. *Free Radic Biol Med* 88, 452–465.

[7] Dean, M., Hamon, Y., and Chimini, G. (2001) The human ATP-binding cassette (ABC) transporter superfamily. *J Lipid Res* 42, 1007–1017.

[8] Zhang, Y., and Gordon, G. B. (2004) A strategy for cancer prevention: stimulation of the Nrf2-ARE signaling pathway. *Mol Cancer Ther* 3, 885–893.

[9] Cooke, M. S., Evans, M. D., Dizdaroglu, M., and Lunec, J. (2003) Oxidative DNA damage: mechanisms, mutation, and disease. *FASEB J* 17, 1195–1214.

[10] Harman, D. (1956) Aging: a theory based on free radical and radiation chemistry. *J Gerontol* 11, 298–300.

[11] Kensler, T. W., and Wakabayashi, N. (2010) Nrf2: friend or foe for chemoprevention? *Carcinogenesis* 31, 90–99.

[12] Pergola, P. E., Raskin, P., Toto, R. D., Meyer, C. J., Huff, J. W., Grossman, E. B., Krauth, M., Ruiz, S., Audhya, P., Christ-Schmidt, H., Wittes, J., Warnock, D. G., and Investigators, B. S. (2011) Bardoxolone methyl and kidney function in CKD with type 2 diabetes. *N Engl J Med* 365, 327–336.

[13] de Zeeuw, D., Akizawa, T., Audhya, P., Bakris, G. L., Chin, M., Christ-Schmidt, H., Goldsberry, A., Houser, M., Krauth, M., Lambers Heerspink, H. J., McMurray, J. J., Meyer, C. J., Parving, H. H., Remuzzi, G., Toto, R. D., Vaziri, N. D., Wanner, C., Wittes, J., Wrolstad, D., Chertow, G. M., and Investigators, B. T. (2013) Bardoxolone methyl in type 2 diabetes and stage 4 chronic kidney disease. *N Engl J Med* 369, 2492–2503.

[14] Doss, J. F., Jonassaint, J. C., Garrett, M. E., Ashley-Koch, A. E., Telen, M. J., and Chi, J. T. (2016) Phase 1 study of a sulforaphane-containing broccoli sprout homogenate for sickle cell disease. *PLoS One* 11, e0152895.

[15] Castro, J. P., Grune, T., and Speckmann, B. (2016) The two faces of reactive oxygen species (ROS) in adipocyte function and dysfunction. *Biol Chem* 397, 709–724.

[16] Luperini, B. C., Almeida, D. C., Porto, M. P., Marcondes, J. P., Prado, R. P., Rasera, I., Oliveira, M. R., and Salvadori, D. M. (2015) Gene polymorphisms and increased DNA damage in morbidly obese women. *Mutat Res* 776, 111–117.

[17] Fernandez-Sanchez, A., Madrigal-Santillan, E., Bautista, M., Esquivel-Soto, J., Morales-Gonzalez, A., Esquivel-Chirino, C., Durante-Montiel, I., Sanchez-Rivera, G., Valadez-Vega, C., and Morales-Gonzalez, J. A. (2011) Inflammation, oxidative stress, and obesity. *Int J Mol Sci* 12, 3117–3132.

[18] Probst, B. L., McCauley, L., Trevino, I., Wigley, W. C., and Ferguson, D. A. (2015) cancer cell growth is differentially affected by constitutive activation of NRF2 by KEAP1 deletion and pharmacological activation of NRF2 by the synthetic triterpenoid, RTA 405. *PLoS One* 10, e0135257.

[19] Furfaro, A. L., Traverso, N., Domenicotti, C., Piras, S., Moretta, L., Marinari, U. M., Pronzato, M. A., and Nitti, M. (2016) The Nrf2/HO-1 axis in cancer cell growth and chemoresistance. *Oxid Med Cell Longev* 2016, 1958174

[20] Hintsala, H. R., Jokinen, E., Haapasaari, K. M., Moza, M., Ristimaki, A., Soini, Y., Koivunen, J., and Karihtala, P. (2016) Nrf2/Keap1 pathway and expression of oxidative stress lesions 8-hydroxy-2?-deoxyguanosine and nitrotyrosine in melanoma. *Anticancer Res* 36, 1497–1506.

[21] Cancer Genome Atlas Research, N. (2014) Comprehensive molecular profiling of lung adenocarcinoma. *Nature* 511, 543–550.

[22] Cancer Genome Atlas Research, N. (2012) Comprehensive genomic characterization of squamous cell lung cancers. *Nature* 489, 519–525.

[23] Yoh, K., Itoh, K., Enomoto, A., Hirayama, A., Yamaguchi, N., Kobayashi, M., Morito, N., Koyama, A., Yamamoto, M., and Takahashi, S. (2001) Nrf2-deficient female mice develop lupus-like autoimmune nephritis. *Kidney Int* 60, 1343–1353.

[24] Lee, J. M., Chan, K., Kan, Y. W., and Johnson, J. A. (2004) Targeted disruption of Nrf2 causes regenerative immune-mediated hemolytic anemia. *Proc Natl Acad Sci U S A* 101, 9751–9756.

[25] Ma, Q., Battelli, L., and Hubbs, A. F. (2006) Multiorgan autoimmune inflammation, enhanced lymphoproliferation, and impaired homeostasis of reactive oxygen species in mice lacking the antioxidant-activated transcription factor Nrf2. *Am J Pathol* 168, 1960–1974.

[26] Cho, H. Y., and Kleeberger, S. R. (2010) Nrf2 protects against airway disorders. *Toxicol Appl Pharmacol* 244, 43–56.

[27] Ryoo, I. G., Ha, H., and Kwak, M. K. (2014) Inhibitory role of the KEAP1-NRF2 pathway in TGFbeta1-stimulated renal epithelial transition to fibroblastic cells: a modulatory effect on SMAD signaling. *PLoS One* 9, e93265.

[28] Ryoo, I. G., Shin, D. H., Kang, K. S., and Kwak, M. K. (2015) Involvement of Nrf2-GSH signaling in TGFbeta1-stimulated epithelial-to-mesenchymal transition changes in rat renal tubular cells. *Arch Pharm Res* 38, 272–281.

[29] Bolati, D., Shimizu, H., Yisireyili, M., Nishijima, F., and Niwa, T. (2013) Indoxyl sulfate, a uremic toxin, downregulates renal expression of Nrf2 through activation of NF-kappaB. *BMC Nephrol* 14, 56.

[30] Cho, H. Y., Jedlicka, A. E., Reddy, S. P., Kensler, T. W., Yamamoto, M., Zhang, L. Y., and Kleeberger, S. R. (2002) Role of NRF2 in protection against hyperoxic lung injury in mice. *Am J Respir Cell Mol Biol* 26, 175–182.

[31] Cho, H. Y., Reddy, S. P., Yamamoto, M., and Kleeberger, S. R. (2004) The transcription factor NRF2 protects against pulmonary fibrosis. *FASEB J* 18, 1258–1260.

[32] Chan, K., and Kan, Y. W. (1999) Nrf2 is essential for protection against acute pulmonary injury in mice. *Proc Natl Acad Sci U S A* 96, 12731–12736.

[33] de la Vega, M. R., Dodson, M., Gross, C., Manzour, H., Lantz, R. C., Chapman, E., Wang, T., Black, S. M., Garcia, J. G., and Zhang, D. D. (2016) Role of Nrf2 and autophagy in acute lung injury. *Curr Pharmacol Rep* 2, 91–101.

[34] Lee, J. M., Li, J., Johnson, D. A., Stein, T. D., Kraft, A. D., Calkins, M. J., Jakel, R. J., and Johnson, J. A. (2005) Nrf2, a multi-organ protector? *FASEB J* 19, 1061–1066.

[35] Ramos-Gomez, M., Dolan, P. M., Itoh, K., Yamamoto, M., and Kensler, T. W. (2003) Interactive effects of nrf2 genotype and oltipraz on benzo[a]pyrene-DNA adducts and tumor yield in mice. *Carcinogenesis* 24, 461–467.

[36] Shelton, P., and Jaiswal, A. K. (2013) The transcription factor NF-E2-related factor 2 (Nrf2): a protooncogene? *FASEB J* 27, 414–423.

[37] Mitsuishi, Y., Taguchi, K., Kawatani, Y., Shibata, T., Nukiwa, T., Aburatani, H., Yamamoto, M., and Motohashi, H. (2012) Nrf2 redirects glucose and glutamine into anabolic pathways in metabolic reprogramming. *Cancer Cell* 22, 66–79.

[38] DeNicola, G. M., Chen, P. H., Mullarky, E., Sudderth, J. A., Hu, Z., Wu, D., Tang, H., Xie, Y., Asara, J. M., Huffman, K. E., Wistuba, II, Minna, J. D., DeBerardinis, R. J., and Cantley, L. C. (2015) NRF2 regulates serine biosynthesis in non-small cell lung cancer. *Nat Genet* 47, 1475–1481.

[39] Menegon, S., Columbano, A., and Giordano, S. (2016) The dual roles of NRF2 in cancer. *Trends Mol Med* 22, 578–593.

[40] Wang, X. J., Sun, Z., Villeneuve, N. F., Zhang, S., Zhao, F., Li, Y., Chen, W., Yi, X., Zheng, W., Wondrak, G. T., Wong, P. K., and Zhang, D. D. (2008) Nrf2 enhances resistance of cancer cells to chemotherapeutic drugs, the dark side of Nrf2. *Carcinogenesis* 29, 1235–1243.

[41] Homma, S., Ishii, Y., Morishima, Y., Yamadori, T., Matsuno, Y., Haraguchi, N., Kikuchi, N., Satoh, H., Sakamoto, T., Hizawa, N., Itoh, K., and Yamamoto, M. (2009) Nrf2 enhances cell proliferation and resistance to anticancer drugs in human lung cancer. *Clin Cancer Res* 15, 3423–3432.

[42] DeNicola, G. M., Karreth, F. A., Humpton, T. J., Gopinathan, A., Wei, C., Frese, K., Mangal, D., Yu, K. H., Yeo, C. J., Calhoun, E. S., Scrimieri, F., Winter, J. M., Hruban, R. H., Iacobuzio-Donahue, C., Kern, S. E., Blair, I. A., and Tuveson, D. A. (2011) Oncogene-induced Nrf2 transcription promotes ROS detoxification and tumorigenesis. *Nature* 475, 106–109.

[43] Bai, X., Chen, Y., Hou, X., Huang, M., and Jin, J. (2016) Emerging role of NRF2 in chemoresistance by regulating drug-metabolizing enzymes and efflux transporters. *Drug Metab Rev*, 1–27.

[44] Qaisiya, M., Coda Zabetta, C. D., Bellarosa, C., and Tiribelli, C. (2014) Bilirubin mediated oxidative stress involves antioxidant response activation via Nrf2 pathway. *Cell Signal* 26, 512–520.

[45] Li, J., Lee, J. M., and Johnson, J. A. (2002) Microarray analysis reveals an antioxidant responsive element-driven gene set involved in conferring protection from an oxidative stress-induced apoptosis in IMR-32 cells. *J Biol Chem* 277, 388–394.

[46] Nault, J. C., Rebouissou, S., and Zucman Rossi, J. (2015) NRF2/KEAP1 and Wnt/beta-catenin in the multistep process of liver carcinogenesis in humans and rats. *Hepatology* 62, 677–679.

[47] Funes, J. M., Henderson, S., Kaufman, R., Flanagan, J. M., Robson, M., Pedley, B., Moncada, S., and Boshoff, C. (2014) Oncogenic transformation of mesenchymal stem cells decreases Nrf2 expression favoring in vivo tumor growth and poorer survival. *Mol Cancer* 13, 20.

[48] Maki, Y., Fujimoto, J., Lang, W., Xu, L., Behrens, C., Wistuba, II, and Kadara, H. (2015) LAPTM4B is associated with poor prognosis in NSCLC and promotes the NRF2-mediated stress response pathway in lung cancer cells. *Sci Rep* 5, 13846.

[49] Kanamori, M., Higa, T., Sonoda, Y., Murakami, S., Dodo, M., Kitamura, H., Taguchi, K., Shibata, T., Watanabe, M., Suzuki, H., Shibahara, I., Saito, R., Yamashita, Y., Kumabe, T., Yamamoto, M., Motohashi, H., and Tominaga, T. (2015) Activation of the NRF2 pathway and its impact on the prognosis of anaplastic glioma patients. *Neuro Oncol* 17, 555–565.

[50] Ji, X., Wang, H., Zhu, J., Zhu, L., Pan, H., Li, W., Zhou, Y., Cong, Z., Yan, F., and Chen, S. (2014) Knockdown of Nrf2 suppresses glioblastoma angiogenesis by inhibiting hypoxia-induced activation of HIF-1alpha. *Int J Cancer* 135, 574–584.

[51] Soini, Y., Eskelinen, M., Juvonen, P., Karja, V., Haapasaari, K. M., Saarela, A., and Karihtala, P. (2014) Nuclear Nrf2 expression is related to a poor survival in pancreatic adenocarcinoma. *Pathol Res Pract* 210, 35–39.

[52] Ko, H., Kim, S. J., Shim, S. H., Chang, H., and Ha, C. H. (2016) Shikonin induces the apoptotic cell death via regulation of p53 and Nrf2 in AGS human stomach carcinoma cells. *Biomol Ther (Seoul)* 2016, 1–9.

[53] Ryoo, I. G., Lee, S. H., and Kwak, M. K. (2016) Redox modulating NRF2: a potential mediator of cancer stem cell resistance. *Oxid Med Cell Longev* 2016, 2428153.

[54] Hu, X. F., Yao, J., Gao, S. G., Wang, X. S., Peng, X. Q., Yang, Y. T., and Feng, X. S. (2013) Nrf2 overexpression predicts prognosis and 5-FU resistance in gastric cancer. *Asian Pac J Cancer Prev* 14, 5231–5235.

[55] Shibata,T.,Kokubu,A.,Gotoh,M.,Ojima,H.,Ohta,T.,Yamamoto,M.,andHirohashi,S.(2008) Genetic alteration of Keap1 confers constitutive Nrf2 activation and resistance to chemotherapyingallbladdercancer.*Gastroenterology*135,1358–1368,1368e1351–e1354.

[56] Hong, Y. B., Kang, H. J., Kwon, S. Y., Kim, H. J., Kwon, K. Y., Cho, C. H., Lee, J. M., Kallakury, B. V., and Bae, I. (2010) Nuclear factor (erythroid-derived 2)-like 2 regulates drug resistance in pancreatic cancer cells. *Pancreas* 39, 463–472.

[57] Wang, H., Liu, X., Long, M., Huang, Y., Zhang, L., Zhang, R., Zheng, Y., Liao, X., Wang, Y., Liao, Q., Li, W., Tang, Z., Tong, Q., Wang, X., Fang, F., Rojo de la Vega, M., Ouyang, Q., Zhang, D. D., Yu, S., and Zheng, H. (2016) NRF2 activation by antioxidant antidiabetic agents accelerates tumor metastasis. *Sci Transl Med* 8, 334–351.

[58] Sporn, M. B., and Liby, K. T. (2012) NRF2 and cancer: the good, the bad and the importance of context. *Nat Rev Cancer* 12, 564–571.

[59] Hast, B. E., Cloer, E. W., Goldfarb, D., Li, H., Siesser, P. F., Yan, F., Walter, V., Zheng, N., Hayes, D. N., and Major, M. B. (2014) Cancer-derived mutations in KEAP1 impair NRF2 degradation but not ubiquitination. *Cancer Res* 74, 808–817.

[60] Hanada, N., Takahata, T., Zhou, Q., Ye, X., Sun, R., Itoh, J., Ishiguro, A., Kijima, H., Mimura, J., Itoh, K., Fukuda, S., and Saijo, Y. (2012) Methylation of the KEAP1 gene promoter region in human colorectal cancer. *BMC Cancer* 12, 66.

[61] Liu, Q. L., Zhang, J., Liu, X., and Gao, J. Y. (2015) Role of growth hormone in maturation and activation of dendritic cells via miR-200a and the Keap1/Nrf2 pathway. *Cell Prolif* 48, 573–581.

[62] Eades, G., Yang, M., Yao, Y., Zhang, Y., and Zhou, Q. (2011) miR-200a regulates Nrf2 activation by targeting Keap1 mRNA in breast cancer cells. *J Biol Chem* 286, 40725–40733.

[63] Shi, L., Wu, L., Chen, Z., Yang, J., Chen, X., Yu, F., Zheng, F., and Lin, X. (2015) MiR-141 activates Nrf2-dependent antioxidant pathway via down-regulating the expression of Keap1 conferring the resistance of hepatocellular carcinoma cells to 5-fluorouracil. *Cell Physiol Biochem* 35, 2333–2348.

[64] Yang,M.,Yao,Y.,Eades,G.,Zhang,Y.,andZhou,Q.(2011)MiR-28regulatesNrf2expression throughaKeap1-independentmechanism.Breast*CancerResTreat*129,983–991.

[65] Khunluck, T., Kukongviriyapan, V., Puapairoj, A., Khuntikeo, N., Senggunprai, L., Zeekpudsa,P.,andPrawan,A.(2014)AssociationofNRF2polymorphismwithcholangiocarcinomaprognosisinThaipatients.*AsianPacJCancerPrev*15,299–304.

[66] Ishikawa, T. (2014) Genetic polymorphism in the NRF2 gene as a prognosis marker for cancer chemotherapy. *Front Genet* 5, 383

[67] Hartikainen, J. M., Tengstrom, M., Kosma, V. M., Kinnula, V. L., Mannermaa, A., and Soini, Y. (2012) Genetic polymorphisms and protein expression of NRF2 and Sulfiredoxin predict survival outcomes in breast cancer. *Cancer Res* 72, 5537–5546.

[68] Cho, H. Y., Jedlicka, A. E., Gladwell, W., Marzec, J., McCaw, Z. R., Bienstock, R. J., and Kleeberger, S. R. (2015) Association of Nrf2 polymorphism haplotypes with acute lung injury phenotypes in inbred strains of mice. *Antioxid Redox Signal* 22, 325–338.

[69] Marzec, J. M., Christie, J. D., Reddy, S. P., Jedlicka, A. E., Vuong, H., Lanken, P. N., Aplenc, R., Yamamoto, T., Yamamoto, M., Cho, H. Y., and Kleeberger, S. R. (2007) Functional polymorphisms in the transcription factor NRF2 in humans increase the risk of acute lung injury. *FASEB J* 21, 2237–2246.

[70] Shimoyama, Y., Mitsuda, Y., Tsuruta, Y., Hamajima, N., and Niwa, T. (2014) Polymorphism of Nrf2, an antioxidative gene, is associated with blood pressure and cardiovascular mortality in hemodialysis patients. *Int J Med Sci* 11, 726–731.

[71] Villeneuve, N. F., Tian, W., Wu, T., Sun, Z., Lau, A., Chapman, E., Fang, D., and Zhang, D. D. (2013) USP15 negatively regulates Nrf2 through deubiquitination of Keap1. *Mol Cell* 51, 68–79.

[72] Arlt, A., Sebens, S., Krebs, S., Geismann, C., Grossmann, M., Kruse, M. L., Schreiber, S., and Schafer, H. (2013) Inhibition of the Nrf2 transcription factor by the alkaloid trigonelline renders pancreatic cancer cells more susceptible to apoptosis through decreased proteasomal gene expression and proteasome activity. *Oncogene* 32, 4825–4835.

[73] May, J. M. (2000) How does ascorbic acid prevent endothelial dysfunction? *Free Radic Biol Med* 28, 1421–1429.

[74] Wagner, A. E., Boesch-Saadatmandi, C., Breckwoldt, D., Schrader, C., Schmelzer, C., Doring, F., Hashida, K., Hori, O., Matsugo, S., and Rimbach, G. (2011) Ascorbic acid partly antagonizes resveratrol mediated heme oxygenase-1 but not paraoxonase-1

induction in cultured hepatocytes—role of the redox-regulated transcription factor Nrf2. *BMC Complement Altern Med* 11, 1.

[75] Messier, E. M., Bahmed, K., Tuder, R. M., Chu, H. W., Bowler, R. P., and Kosmider, B. (2013) Trolox contributes to Nrf2-mediated protection of human and murine primary alveolar type II cells from injury by cigarette smoke. *Cell Death Dis* 4, e573

[76] Aon, M. A., Cortassa, S., and O'Rourke, B. (2010) Redox-optimized ROS balance: a unifying hypothesis. *Biochim Biophys Acta* 1797, 865–877.

[77] Sun, S. Y. (2010) N-acetylcysteine, reactive oxygen species and beyond. *Cancer Biol Ther* 9, 109–110.

[78] Preston, T. J., Muller, W. J., and Singh, G. (2001) Scavenging of extracellular H_2O_2 by catalase inhibits the proliferation of HER-2/Neu-transformed rat-1 fibroblasts through the induction of a stress response. *J Biol Chem* 276, 9558–9564.

[79] Olayanju, A., Copple, I. M., Bryan, H. K., Edge, G. T., Sison, R. L., Wong, M. W., Lai, Z. Q., Lin, Z. X., Dunn, K., Sanderson, C. M., Alghanem, A. F., Cross, M. J., Ellis, E. C., Ingelman-Sundberg, M., Malik, H. Z., Kitteringham, N. R., Goldring, C. E., and Park, B. K. (2015) Brusatol provokes a rapid and transient inhibition of Nrf2 signaling and sensitizes mammalian cells to chemical toxicity-implications for therapeutic targeting of Nrf2. *Free Radic Biol Med* 78, 202–212.

[80] Sun, X., Wang, Q., Wang, Y., Du, L., Xu, C., and Liu, Q. (2016) Brusatol enhances the radiosensitivity of A549 cells by promoting ROS production and enhancing DNA damage. *Int J Mol Sci* 17, 997–1010.

[81] Ren, D., Villeneuve, N. F., Jiang, T., Wu, T., Lau, A., Toppin, H. A., and Zhang, D. D. (2011) Brusatol enhances the efficacy of chemotherapy by inhibiting the Nrf2-mediated defense mechanism. *Proc Natl Acad Sci U S A* 108, 1433–1438.

[82] Salminen, A., Kaarniranta, K., and Kauppinen, A. (2016) AMPK and HIF signaling pathways regulate both longevity and cancer growth: the good news and the bad news about survival mechanisms. *Biogerontology* 17, 655–680.

[83] Lee, S. L., Ryu, H., Son, A. R., Seo, B., Kim, J., Jung, S. Y., Song, J. Y., Hwang, S. G., and Ahn, J. (2016) TGF-beta and hypoxia/reoxygenation promote radioresistance of A549 lung cancer cells through activation of Nrf2 and EGFR. *Oxid Med Cell Longev* 2016, 6823471

[84] Levonen, A. L. (2014) Activation of stress signaling pathways by oxidized and nitrated lipids. *Free Radic Biol Med* 75 Suppl 1, S8

[85] Kansanen, E., Jyrkkanen, H. K., and Levonen, A. L. (2012) Activation of stress signaling pathways by electrophilic oxidized and nitrated lipids. *Free Radic Biol Med* 52, 973–982.

[86] Taguchi, K., Motohashi, H., and Yamamoto, M. (2011) Molecular mechanisms of the Keap1-Nrf2 pathway in stress response and cancer evolution. *Genes Cells* 16, 123–140.

[87] Miao, W., Hu, L., Scrivens, P. J., and Batist, G. (2005) Transcriptional regulation of NF-E2 p45-related factor (NRF2) expression by the aryl hydrocarbon receptor-xenobiotic response element signaling pathway: direct cross-talk between phase I and II drug-metabolizing enzymes. *J Biol Chem* 280, 20340–20348.

[88] Yeager, R. L., Reisman, S. A., Aleksunes, L. M., and Klaassen, C. D. (2009) Introducing the "TCDD-inducible AhR-Nrf2 gene battery". *Toxicol Sci* 111, 238–246.

[89] Lu, H., Cui, W., and Klaassen, C. D. (2011) Nrf2 protects against 2,3,7,8-tetrachlorodibenzo-p-dioxin (TCDD)-induced oxidative injury and steatohepatitis. *Toxicol Appl Pharmacol* 256, 122–135.

[90] Kalthoff, S., Ehmer, U., Freiberg, N., Manns, M. P., and Strassburg, C. P. (2010) Interaction between oxidative stress sensor Nrf2 and xenobiotic-activated aryl hydrocarbon receptor in the regulation of the human phase II detoxifying UDP-glucuronosyltransferase 1A10. *J Biol Chem* 285, 5993–6002.

[91] Shin, S., Wakabayashi, N., Misra, V., Biswal, S., Lee, G. H., Agoston, E. S., Yamamoto, M., and Kensler, T. W. (2007) NRF2 modulates aryl hydrocarbon receptor signaling: influence on adipogenesis. *Mol Cell Biol* 27, 7188–7197.

[92] Kim, S. J., Park, C., Han, A. L., Youn, M. J., Lee, J. H., Kim, Y., Kim, E. S., Kim, H. J., Kim, J. K., Lee, H. K., Chung, S. Y., So, H., and Park, R. (2009) Ebselen attenuates cisplatin-induced ROS generation through Nrf2 activation in auditory cells. *Hear Res* 251, 70–82.

[93] Zhang, Y., Talalay, P., Cho, C. G., and Posner, G. H. (1992) A major inducer of anticarcinogenic protective enzymes from broccoli: isolation and elucidation of structure. *Proc Natl Acad Sci U S A* 89, 2399–2403.

[94] Wells, G. (2015) Peptide and small molecule inhibitors of the Keap1-Nrf2 protein-protein interaction. *Biochem Soc Trans* 43, 674–679.

[95] Ammon,H.P.,andWahl,M.A.(1991)Pharmacologyof*Curcumalonga.PlantaMed*57,1–7.

[96] Talero, E., Avila-Roman, J., and Motilva, V. (2012) Chemoprevention with phytonutrients and microalgae products in chronic inflammation and colon cancer. *Curr Pharm Des* 18, 3939–3965.

[97] Khan, S., Karmokar, A., Howells, L., Thomas, A. L., Bayliss, R., Gescher, A., and Brown, K. (2016) Targeting cancer stem-like cells using dietary-derived agents—where are we now? *Mol Nutr Food Res* 60, 1295–1309.

[98] Gao, S., Duan, X., Wang, X., Dong, D., Liu, D., Li, X., Sun, G., and Li, B. (2013) Curcumin attenuates arsenic-induced hepatic injuries and oxidative stress in experimental mice through activation of Nrf2 pathway, promotion of arsenic methylation and urinary excretion. *Food Chem Toxicol* 59, 739–747.

[99] Balogun, E., Hoque, M., Gong, P., Killeen, E., Green, C. J., Foresti, R., Alam, J., and Motterlini, R. (2003) Curcumin activates the haem oxygenase-1 gene via

regulation of Nrf2 and the antioxidant-responsive element. *Biochem J* 371, 887–895.

[100] Bonnefont-Rousselot, D. (2016) Resveratrol and cardiovascular diseases. *Nutrients* 8, 250–274.

[101] Kalthoff, S., Ehmer, U., Freiberg, N., Manns, M. P., and Strassburg, C. P. (2010) Coffee induces expression of glucuronosyltransferases by the aryl hydrocarbon receptor and Nrf2 in liver and stomach. *Gastroenterology* 139, 1699–1710.

[102] Li, M., Wang, X. F., Shi, J. J., Li, Y. P., Yang, N., Zhai, S., and Dang, S. S. (2015) Caffeic acid phenethyl ester inhibits liver fibrosis in rats. *World J Gastroenterol* 21, 3893–3903.

[103] Huang, T. C., Chung, Y. L., Wu, M. L., and Chuang, S. M. (2011) Cinnamaldehyde enhances Nrf2 nuclear translocation to upregulate phase II detoxifying enzyme expression in HepG2 cells. *J Agric Food Chem* 59, 5164–5171.

[104] Wang, F., Pu, C., Zhou, P., Wang, P., Liang, D., Wang, Q., Hu, Y., Li, B., and Hao, X. (2015) Cinnamaldehyde prevents endothelial dysfunction induced by high glucose by activating Nrf2. *Cell Physiol Biochem* 36, 315–324.

[105] Huang, C. S., Lii, C. K., Lin, A. H., Yeh, Y. W., Yao, H. T., Li, C. C., Wang, T. S., and Chen, H. W. (2013) Protection by chrysin, apigenin, and luteolin against oxidative stress is mediated by the Nrf2-dependent up-regulation of heme oxygenase 1 and glutamate cysteine ligase in rat primary hepatocytes. *Arch Toxicol* 87, 167–178.

[106] Miao, W., Hu, L., Kandouz, M., and Batist, G. (2003) Oltipraz is a bifunctional inducer activating both phase I and phase II drug-metabolizing enzymes via the xenobiotic responsive element. *Mol Pharmacol* 64, 346–354.

[107] Wajda, A., Lapczuk, J., Grabowska, M., Slojewski, M., Laszczynska, M., Urasinska, E., and Drozdzik, M. (2016) Nuclear factor E2-related factor-2 (Nrf2) expression and regulation in male reproductive tract. *Pharmacol Rep* 68, 101–108.

[108] Rao, J., Qian, X., Li, G., Pan, X., Zhang, C., Zhang, F., Zhai, Y., Wang, X., and Lu, L. (2015) ATF3-mediated NRF2/HO-1 signaling regulates TLR4 innate immune responses in mouse liver ischemia/reperfusion injury. *Am J Transplant* 15, 76–87.

[109] Imhoff, B. R., and Hansen, J. M. (2010) Tert-butylhydroquinone induces mitochondrial oxidative stress causing Nrf2 activation. *Cell Biol Toxicol* 26, 541–551.

[110] Shi, X., Li, Y., Hu, J., and Yu, B. (2016) Tert-butylhydroquinone attenuates the ethanol-induced apoptosis of and activates the Nrf2 antioxidant defense pathway in H9c2 cardiomyocytes. *Int J Mol Med* 38, 123–130.

[111] No, J. H., Kim, Y. B., and Song, Y. S. (2014) Targeting nrf2 signaling to combat chemoresistance. *J Cancer Prev* 19, 111–117.

Chapter 6

Possible Role of Nrf2 in Oxidative and Inflammatory Processes During Menopause

Claudia C. Calzada Mendoza,
Carlos A. Jiménez Zamarripa,
María E. Ocharán Hernández and
Gisela Gutiérrez Iglesias

Additional information is available at the end of the chapter

http://dx.doi.org/10.5772/65244

Abstract

The increase in life expectancy leads to the possibility of development chronic diseases, from special physiological conditions as occurs in the menopause, which is defined as the permanent cessation of ovulation, marked by the end of menstruation. It has been related to decreased ovarian function that occurs around an age of 45 years. This event involves the reduction in estrogen production and may contribute to the development of chronic-degenerative diseases. Many diseases developed during menopause have been associated with oxidative stress, such as osteoporosis, hot flushes, cognitive impairment, insulin resistance, dry skin, obesity, and cardiovascular events. The knowledge about the participation of Nrf2 in diseases that occur during menopause is very limited. Here, only diseases such as osteoporosis, cardiovascular diseases and dry skin, which are present during menopause and its later stages have been described. The Nrf2 pathway involves the participation of PI3K/Akt, MAPK, and eNOS, which act as mediators for cytoprotection and antioxidation. Compounds such as equol, fitoestrogens, alkyl cathecols, or curcumin could be offered as options to antioxidant treatment, added the fact that they are present in fruits and vegetables which are rich in vitamins, minerals and calcium, thus including all the required nutrients for an adequate nutrition.

Keywords: menopause, cardiovascular diseases, osteoporosis, obesity, inflammation, Nrf2

1. Introduction

Menopause is defined as the permanent cessation of menstruation by an year in a row, by decreasing the estrogen production, as a consequence of ovarian dysfunction. It occurs between 47 and 55 years of age and during the transition, women experience multiple symptoms [1].

Postmenopause is divided into two periods; an early stage, which includes the first five years after confirmation, and a later phase, which starts five years after the onset of menopause and lasts until the end of life; although the latter overlaps with aging. It is important to consider stage of menopause the woman is, because the metabolic disorders begin and are susceptible to be modified, also the biological functions change with age, environment and the co-morbidities.

The end of the reproductive phase and the onset of menopause brings several metabolic disorders, such as frequent vaginal infections, dyslipidemia, weight gain, visceral obesity, hyperinsulinemia, insulin resistance, osteoporosis, glucose impaired tolerance, mild cognitive impairment, alter coagulation, atherosclerosis, hardening of arteries, hypertension, dry skin and mucous (burning mouth syndrome); in addition, night sweats, and mood changes [2].

Several of the aforementioned disorders have been linked to oxidative stress and inflammation. In recent decades, it has been recognized that oxidative stress causes aging and several pathological conditions. In this regard, the menopause has both conditions, the susceptibility to the development diseases related to oxidative stress and aging.

Central adiposity has been linked to insulin resistance (measured using HOMA: homeostasis model assessment) and oxidative stress (oxidized low-density lipoprotein, urinary isoprostanes [PGF2a], protein carbonyls, and DNA damage), coupled with the transient accumulation of iron in postmenopausal women, which provide ideal conditions for an inflammatory and oxidant state, which all together increases cardiovascular risk. So reducing centralized fat mass and maintaining a favorable lipid profile, antioxidant status and iron status may be important in protecting postmenopausal women from atherosclerotic disease [3].

Coronary artery disease, has a higher prevalence in obese postmenopausal women (with high levels of malondialdehyde and lower superoxide dismutase), but is less observed in younger women [4, 5].

Arterial stiffening worsens across the stages of the menopausal transition, which seems to be mediated, by oxidative stress, particularly during the late perimenopausal and postmenopausal periods. Through a model of rapid arterial dilating by infusion of ascorbic acid, it was found that the postmenopausal women have minor vasodilation and higher oxidative stress [6].

Considering the significance of oxidative stress in several diseases, the research studies have been focused at nuclear factor of transcription, Nrf2; which is considered as the most important regulator of the antioxidant response. Nrf2 modulates expression of many genes (related to antioxidant enzymes, inflammatory processes, tissue remodeling, carcinogenesis, and cognitive impairment). The participation of NrF2 in menopausal events has been studied in relationship to mainly, the vascular system, components of metabolic syndrome, osteoporosis, and skin.

Cardiovascular system: Nrf2 is an important component in antioxidant defenses in cardiovascular diseases, such as atherosclerosis, hypertension, and heart failure. Nrf2 is also involved in protection against oxidant stress during the processes of ischemia-reperfusion injury and aging. However, evidence suggests that Nrf2 activity can attenuate or stimulate cardiovascular disease processes.

Oxidative stress is an important factor to the development of endothelial dysfunction, and it has described that the polymorphisms 653A/G (rs35652124), −651G/A (rs6706649), and −617C/A (rs6721961) are located in the promoter region of the gene encoding NRF2 (NFE2L2) and can participate in forearm blood flow (FBF) and forearm vascular resistance (FVR) depending on the ethnic group. For instance, in African Americans −653G variant allele carriers had significantly lower FBF and higher FVR. In other hand, in White Americans, −617A variant allele carriers had significantly higher FVR. Polymorphisms within the NFE2L2 promoter were associated with impaired forearm vasodilator responses in an endothelial-independent manner, suggesting an important role of NRF2 in the regulation of vascular function in humans [7].

Aldosterone activates and increases Nrf2, this effect depends on the mineralocorticoid receptor and oxidative stress. In vivo, Nrf2 activation has beneficial effects on high blood pressure caused by aldosterone [8].

Skin: It has been observed that skin of women turns dry, during menopause, which is also exposed to the reactive oxygen species generated during cell metabolism or by accumulation of fat below skin. It is known that skin cells expresses the transcription factor Nrf2, before menopause, but its expression in postmenopausal stage is unclear [9].

Osteoporosis: This is frequent in postmenopausal women. Recently findings show that this is caused by redox imbalance, even the bone marrow presents higher levels of markers of oxidative stress. An association between elevated hydroperoxides serum levels and reduced bone density in postmenopausal women [10] has been reported.

This chapter presents evidence for the expression of NrF2 in alterations or diseases in menopause associated with oxidative stress.

2. Metabolic disorders during menopause

Menopause is the cessation of the ovarian cycle, so it terminate ovulation and the reproductive stage ends in women. This is because over the years, total ovogonias are reduced in the ovary and most become refractory to the action of the pituitary gonadotropins. As a result, the levels of estradiol are decreased; at the beginning of menopause, the menstruation become irregular and then disappears. The symptoms of menopause are different, depending on the age, culture, preexisting morbidities, diet, and ethnicity [11].

Hormone depletion can increase the vulnerability of tissues that are estrogen-sensitive to the development of diseases. The principle symptoms of menopause include vasomotor symptoms such as night sweats, urogenital atrophy, osteopenia, and osteoporosis, psychiatric disorders, sexual dysfunction, dryness of skin and mucous, cardiovascular disease, cancer, and obesity [2]. Other symptoms are polyuria, fatigue, weakness, irritability, blurred vision, thirst, and increased appetite [12].

The most common metabolic disorders are dyslipidemia, glucose intolerance, insulin resistance, hyperinsulinemia, and type 2 diabetes [13].

The lipid metabolism disorder is characterized by high levels of low density lipoprotein (LDL) and low levels of high-density lipoprotein (HDL) [14]. This disorder makes the woman vulnerable to cardiovascular diseases (CVD) [15]. This is due to the fact that LDL is more susceptible to oxidation reaction, and thus are captured by the macrophage, which triggers an inflammatory process that favors the formation of the atheromatous plaque. Moreover, HDL has the opposite effect, but is diminished in menopause, and the protective effect is lost. Same situation is observed in induced menopause, after a bilateral oophorectomy [16].

Another altered parameter is the plasma glucose concentration. It has been reported that a high incidence of insulin resistance is a risk factor for developing diabetes. The other factors that cause type 2 diabetes are obesity, sedentary lifestyle, poor eating habits, smoking, and alcoholism [17].

After menopause, the incidence of obesity increases, even the body composition changes, from ginecoide type (accumulation in hips and thighs), passes to android type (deposit trunk). Although if the premenopausal women has android-type obesity, they have the same levels of triglycerides and insulin, and risk of metabolic syndrome [18].

The effect of polycystic ovary (PO) syndrome on menopause disorders is controversial. There are studies that support it as an additional cardiovascular risk even during premenopausal phase, because these patients develop early arterial disease, and have a higher prevalence of hypertension, dyslipidemia, incidence of myocardial infarction, thickness of the intima-media, arterial stiffness, and endothelial dysfunction [19]. This is aggravated if a woman has diabetes and dyslipidemia [20]. On the other hand, another study indicates that these patients had fewer climacteric symptoms than controls [21], further have insulin resistance attenuated. But, there were reports that did not find differences between postmenopausal patients with or without pre-polycystic ovary syndrome [22].

Adipocytes (fat cells) secrete leptin, adiponectin, resistin, and ghrelin; the interaction between them modulates the energy balance, appetite, insulin sensitivity, number and size of adipocytes, among other actions that result in the metabolism of fat tissue and the production itself. Additionally, the adipose tissue can synthesize androgens and estrogens, proinflammatory cytokines that may have effects on blood pressure, inflammation process, and lipoprotein metabolism [23].

It has been determined that postmenopausal women with metabolic syndrome have higher levels of serum testosterone levels and protein binding steroid hormones (SHBG)], leptin, resistin, insulin, and HOMA index and low levels of adiponectin. Additionally, the presence of higher level of interleukin-6 (IL-6), and lower level of urokinase plasminogen activator (uPA) were also documented [5]. A side effect of metabolic syndrome is the female sexual dysfunction, in both pre- and post-menopausal women [24].

http://dx.doi.org/10.5772/65244

3. Estrogens and Nrf2

Most disorders and diseases developed during menopause are related to hormonal depletion and the increase in oxidative stress, so that it is easy to assume that estrogens have antioxidant properties. 17β-estradiol is the more potent estrogen, which has been described as capable to inhibit the lipoperoxidation in brain homogenates in rat, induced by Fe^{3+}, due to its lipophilicity and polycyclic groups [25]. Also, this hormone induces the expression of superoxide dismutase, glutathione peroxidase and glutathione S-transferase in peripheral blood mononuclear cells in women who underwent total hysterectomy with bilateral salpingo-oophorectomy and treated with estrogens as HRT [26].

Estrogens have many biological effects, so it is interesting to study compounds of dietary origin with a similar effect. Recently, analysis of molecules similar to estrogens, which are present in vegetables, legumes or fruits, has gained importance due to its possible antioxidant properties as in the case of equol, which is a metabolite of genistein and daidzein, both present in broccoli. The equol has two isomers, the form S-(−)equol and R-(+)equol, of which the first is the most active, but only 30–50% of the population is capable to produce it. The isomer S-(−) can bind to estrogen receptor beta, inhibit MEK, activate eNOS and AMPK and act as antioxidant. It has been observed that this molecule releases nitric oxide, activates Nrf2 and as a consequence promotes the expression of antioxidant genes, before being mediated by the PI3K/Akt pathway, which has been implicated in protection against cytotoxicity, endothelial dysfunction mediated oxidative stress.

In HUVEC cells, S-(−) equol is capable to bind to the membrane estrogen receptor (GPR30) and to activate Nrf2and eNOS through Akt [27].

4. Proxidant conditions during menopause

Iron accumulation: The cessation of menses contributes to accumulation of iron in the body, resulting in elevated serum ferritin. In postmenopausal women, the consequences of this fact are controversial. For example, some studies have reported a direct relationship between iron and atherosclerosis, cerebrovascular risk, oxidation of low density lipoproteins, high cholesterol, inflammation, insulin resistance, and metabolic syndrome [28].

Changes of corporal composition: Greater waist circumference was associated with high oxLDL which is independent of BMI, suggesting that the same abdominal fat mass may induce oxidative stress, more than the general mass fat. Another study reported that the waist/hip ratio is directly associated with LDL-C and lipid oxidation, and inversely with HDL-C, and protein carbonyls or 8-OHdG, injury indicators [29, 30].

Studies carried out in pre and postmenopausal women, found that those who are obese have a higher concentration of malondialdehyde; but the concentration of superoxide dismutase enzyme was similar in all of cases [4].

5. Nrf2 and pathologies related to oxidative stress

5.1. Osteoporosis

Epidemiological studies show that the majority of postmenopausal women are affected by skeletal fragility caused by an excessive bone resorption [24]. Earlier, it was closely related to oxidative stress. It is due to reduction of estrogen plasma levels and its antioxidant action on bone [9, 31].

The estrogens activate Nrf2, which regulates the expression of antioxidant enzymes in bone marrow. As a consequence of this reduction, Nrf2 decreases and leads to the activation of the receptor activator of nuclear factor κB (RANKL), which promotes osteoclastogenesis [32, 33]. Moreover, the Nrf2 deficiency induces a reduction in the ratio cortical area/total area, higher trabecular spacing, osteoclast surface in ovariectomiced mice and Nrf−/− mice [34].

5.2. Cardiovascular diseases

In humans, it has been described, three single nucleotide polymorphisms within the promoter region of the gene encodes NRF2 (NFE2L2) (−653A/G, −651G/A, and −617C/A). The −617A variant allele has major risk to develop lung injury, mediated by oxidative stress. The NFE2L2 polymorphism has been associated to diseases with oxidative stress and inflammation, both mediated by drugs [35].

A study showed that Africans Americans carriers of −653G variant allele, have lower forearm blood flow and higher forearm vascular, resistant; in contrast with White Americans. With respect to −651G/A there were no differences between two populations. In White Americans, the −617 polymorphism carriers have lower forearm blood flow (FBF) and higher forearm vascular resistance (FVR) [6]. In clinical studies, it has been observed that African Americans have a higher prevalence of hypertension and the decreased endothelium compared to the white population. This predisposition to increased vascular resistance in African Americans leads to increased shear stress on endothelial cells, which activates NAD(P)H oxidase and mediates the formation of the oxygen species in the vascular system [36]. Accordingly, the NADPH oxidase equilibrium nitric/superoxide/peroxynitrite oxide in endothelial cells is shifted in favor of the reactive oxygen species in African Americans, which is associated with a deterioration of the vasodilator capacity [37].

The negative effects of pollution, promotes an oxidative state, which is associated with cardiopulmonary diseases. The solid matter alters lung, cardiovascular, nervous system functions, resulting in vascular inflammation, vascular dysfunction, and increased oxidative stress [38–41]. Nrf2 participates in resistance to hyperoxia-induced lung injury, it is also important in the response of epithelial cells particle exposure. The data suggest an alteration in the autonomic regulation of cardiac function during hyperoxia, which is modulated by Nrf2. Therefore, these changes may have important implications FVR for susceptibility to adverse cardiac response outcomes during oxidant exposure [42].

Oxidative stress is a component of the pathogenesis in many cardiovascular diseases, and atherosclerosis, hypertension, heart collapse, and ischemia/reperfusion injury. The sources of

http://dx.doi.org/10.5772/65244

reactive oxygen species (ROS) that lead to oxidative stress due to inefficiencies in the chain of mitochondrial electron transport, NADPH oxidase and xanthine oxidase everywhere, and the metal ions released during cell lysis [43–45].

Atherosclerosis is an inflammatory disease characterized by endothelial filtration and accumulation of oxidized lipoproteins of low density (LDL), physical damage to the endothelium (e.g., turbulent flow of blood, hypertension and/or smoking).

Susceptibility to atheroma formation is not uniform throughout the vascular system, which can be generated by shear stress generated by oscillatory flow, not unidirectional, and turbulent blood flow, which occurs, for example, at junctions or vessel branching points, which are the most susceptible. Conversely, atheromas are less likely to form in the vascular regions with unidirectional laminar blood flow. It is known, that shear-stress laminar flow into blood vessels stimulates the release of nitric (NO) oxide, but when the flow is oscillatory, the stenosis or branch vessels are developed. This reduces NO production and increases superoxide release, leading to oxidative stress and the progression of atherosclerosis. Laminar flow promotes activation of Nrf2, and the oscillatory blood flow suppresses activation of Nrf2, resulting in a favorable environment for atherogenesis [46].

It is increasingly evident that Nrf2 is important for long-term vascular integrity and endothelial functioning, for example, sustained release of NO and protection from apoptosis [47].

The levels of matrix metalloproteinase 9 (MMP9) are linked to plaque destabilization, which produces acute constriction of blood vessel flow and sudden cardiac or events. The atheroprotective effect of HO-1 may be associated with the partial deletion MMP9 to maintain or improve the stability of the plate, avoiding a coronary event or acute and potentially fatal brain [48, 49].

However, the effect of deficiency of Nrf2 on atheromatous plaque are controversial; for example, a study indicates that deficiency of ApoE and Nrf2 (*ApoE−/−Nrf2−/−*)in mice fed with fat diet had minor area of atheromatos plaque and lower softening arterial [50, 51].

Ischemia-reperfusion has been observed in processes such as thrombosis or vasospasm, leading to inflammatory and oxidizing conditions [52]. These conditions promote expression of the Nrf2 oxidant, as evidenced in cell culture cardiomyocytes rat after ischemia-reperfusion cycles, that increased mRNA and protein mRNA of Nrf2. If ischemia is lower, then Nrf2, can attenuate oxidative stress.

During an increase in blood pressure, the renin-angiotensin system increases the concentration of free radicals, also they contribute to this and NADPH oxidases, NOX 1 are known to activate Nrf2 [53].

It has been documented that the diseased myocardium increases oxidative stress, may increase susceptibility to arrhythmia by a direct toxic effect of increased necrosis and apoptosis. In a model overload it was found that the increased expression of Nrf2 decreased myocardial hypertrophy, cardiac fibroblasts and ROS production, the latter, most likely by modulating the activity of Nox4 [54]. Although the sharp increase in the Nrf2 content is cardioprotective, to modulate the production of ROS, apparently chronic activation may have a contrary effect called "reductive stress", so more studies on their impact is required.

ERK's role is controversial, on the one hand it has been reported that oxidative stress index expression of ERK, and participates in Nrf2 signaling, but studies indicate that activation of ERK leads to apoptosis. Studies in rats and monkeys shown the reduction of the expression of several components of the signaling pathway Nrf2 [55–57].

5.3. Skin

Estrogens modulates several actions in skin, in example, they promotes the keratinocyte proliferation, the I and II collagen expression, and represents an antioxidant defense; but when menopause begins, the skin suffer many changes, principally, dryness, becomes thinner, decreased its elasticity, the collagen content is lower, and the vascularity is diminished [58]. Cells are exposed to physical, chemical, mechanical and thermal factors, which can develop oxidative stress, leading to changes in skin appearance, modifying cells and potential malignancies [8].

Nrf2 is expressed in all cell lines of the epidermis. It binds to promoter regions known as elements of antioxidant response (ARE) that transcribed genes of cytoprotective proteins and antioxidants like glutatión S-transferase (GST), quinone reductase NAD(P)H (NQO1), UDP-glucuronosyltransferases (UGT), epoxide hydrolase (EPHX), c glutamylcysteine ligase (GCL), heme oxygenase-1 (HO-1), glutathione reductase the (GR), thioredoxin reductase (TrxR), catalase (CAT) and superoxide dismutase (SOD). Nrf2 also promotes gene transcription nonenzymatic antioxidant proteins as thioredoxin and ferritin [59].

It has been reported that low levels of ROS cause an increase in Nrf2 expression but high levels, do not modify it, this leads to irreversible cell injury and apoptosis induction. An intermediate level of ROS maintains control of the balance between survival and apoptosis by activating transcription factor-NFjB [60].

The epidermis (outer skin layer) consists mainly of keratinocytes and melanocytes, Langerhans cells and Merkel cells. Nrf2 has two effects, firstly, promotes regeneration and expression of antioxidant enzymes [61], and secondly, it can perpetuate the survival of damaged cells by ROS generated during aerobic metabolism, and the respiratory burst of the immune cell system, increasing the resistance to apoptosis of cells in culture under UV ray [62]. Nrf2 can be activated by the sulforane (obtained from broccoli and Brussels sprouts) [8].

In melanocytes, the H_2O_2 is generated by tyrosinase enzyme and later is metabolized by catalase. The Nrf2 activity protects melanocytes against the effects of ROS. This effect is associated with a higher level of melanotropine (a-MSH), this unbalanced signal redox state arrives from the cytoplasm and begins the synthesis of molecules Nrf2 [63, 64].

The synthesis of Nrf2 is constant in melanocytes and alterations in this process, can reduce the resistance of cells to stress, both physical and chemical, leading to cell death or carcinogenesis [65, 66]. The Nrf2 level may be affected by hormones like estrogens. Curcumin is an extract from turmeric root, which acts as antifibrotic in systemic scleroderma, by reducing ROS that causes suppression of the fibrotic process in scleroderma [67, 68]. In contrast, it reduces Nrf2 activity also by antioxidants thiol and thioredoxin, since the thiol group can prevent oxidation of Keap1, which favors the maintenance of Nrf2 in theKeap1 complex [69].

5.4. Immunological disorders

The adipose tissue is an important endocrine gland, it produces proinflammatory cytokines. In menopause, several inflammatory conditions, such as preexist diseases and oxidative stress, have been described, but possibly the most important factor is the centralized fat mass. In a study that considered healthy postmenopausal women of different ethnicity higher levels of proinflammatory markers, such as reactive C-reactive protein, interleukin-1β, and tumor necrosis factorα, are reported. This fact is very important, because, it marked a proinflammatory state associated only with the postmenopause [70].

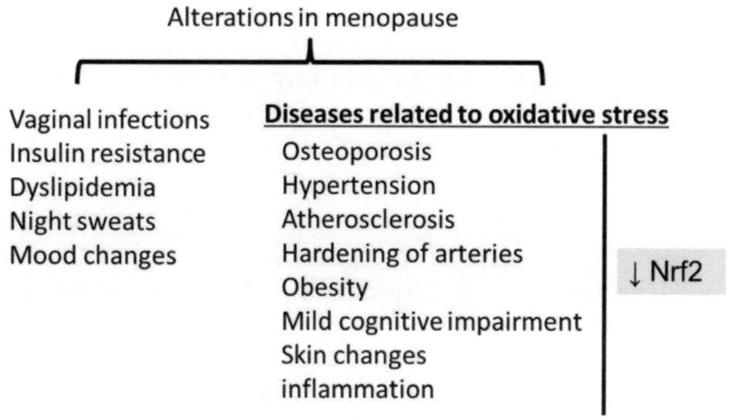

Figure 1. Alterations and diseases observed in menopause. Several diseases present in the menopause, are related to low levels.

The Nrf2 pathway plays a role in degenerative and immune disorders, such as atherosclerosis, inflammatory bowel disease (IBD), diabetes, rheumatoid arthritis, HIV/acquired immunodeficiency (HIV/AIDS) syndrome, neurological disorders, sepsis, cancer, Alzheimer's disease and many others [71]. Although the exact mechanism is unknown, studies in macrophages, indicate that genes Nrf2 inhibits pro-inflammatory cytokines [72], so this mechanism must be deregulated in this stage of life. Many studies on menopause are presented, but there some of them has been related to oxidative stress, which are characterized by a decreased level of Nrf2 (**Figure 1**).

6. Diet and Nrf2

Changes in body composition of women may have different causes, for example, mood swings can lead to eating disorders. Although, special diets are recommended for this stage; investigations are providing substantial information based on its recommendation in relation to the activation of Nrf2 that can induce phytochemicals present in different foods, especially of plant origin. Among these compounds are curcumin, sulforaphane, anthocyanins and alkyl catechol. Lactobacilli are important too, as they contain phenolic acid decarboxylase enzyme, alkyl catechol (**Figure 2**) [73–75].

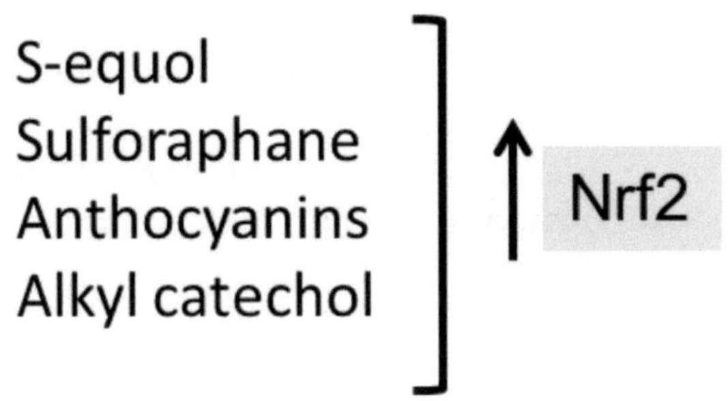

Figure 2. Modulators of Nrf2. Levels of Nrf2 can increase by compounds present in vegetables such as curcumin and broccoli.

7. Conclusions

Menopause is caused by the changes in the production of estrogens, hormones that modulate several functions in organs and tissues, but when its blood levels are diminished, different disorders are developed, such as, cardiovascular events, osteoporosis, dyslipidemia, changes in corporal composition, among others. All of them have been directly related to oxidative stress. The major regulator of this state is Nrf2, which is activated by estrogens, phytoestrogens, and alkyl catechol. In conclusion, the oxidative stress observed during menopause and its stages, can be modulated by activators of Nrf2, and vegetables and fruits have compounds with similar effect.

Author details

Claudia C. Calzada Mendoza[1*], Carlos A. Jiménez Zamarripa[2],
María E. Ocharán Hernández[1] and Gisela Gutiérrez Iglesias[1]

*Address all correspondence to: cccalzadam@yahoo.com.mx

1 Instituto Politécnico Nacional—Escuela Superior de Medicina, Delegation Miguel Hidalgo, Mexico City, Mexico

2 Instituto Politécnico Nacional—Centro Interdisciplinario de Ciencias de la Salud, Mexico City, Mexico

References

[1] Sharma TP, Nett TM, Karsch FJ, Phillips DJ, Lee JS, Herkimer C, Padmanabhan V. Neuroendocrine control of FSH secretion: IV. Hypothalamic control of pituitary FSH-regulatory proteins and their relationship to changes in FSH synthesis and secretion. Biol Reprod. 2012;86(6):171–179. DOI:10.1095/biolreprod.111.098442.

[2] Cray LA, Fugate WN, Herting JR, Sullivan ME. Symptom clusters during the late reproductive stage through the early postmenopause: observations from the Seattle Midlife Women's Health Study. Menopause. 2012 August;19(8):864–869. DOI:10.1097/gme.0b013e31824790a6.

[3] Kyung PJ, Kim M, Kim M, Yen Y, YHA, Sang-Hyun Lee SH, Ho LS. Circulating Lp-PLA2 activity correlates with oxidative stress and cytokines in overweight/obese postmenopausal women not using hormone replacement therapy. Age. 2015;37:32–41. DOI:10.1007/s11357-015-9770-4.

[4] Uppoor RB, Rajesh A, Srinivasan MP, Unnikrishnan B, Holla R. Oxidative stress in obese postmenopausal women: an additive burden for atherosclerosis. J Clin Diagn Res. 2015;9(12):OC03–OC05. DOI:10.7860/JCDR/2015/16467.6868.

[5] Chedraui P, Escobar GS, Pérez-López, Palla G, Montt-Guevara M, Cecchi E, Genazzani AR, Simoncini T. Research group for the Omega Women's Health Project. Angiogenesis, inflammation and endothelial function in postmenopausal women screened for the metabolic syndrome. Maturitas 2014;77:370–374. DOI:10.1016/j.maturitas.2014.01.014.

[6] Hildreth KL, Kohrt WM, Moreau KL. Oxidative stress contributes to large elastic arterial stiffening across the stages of the menopausal transition. Menopause. 2014;21(6):624–632. DOI:10.1097/GME.0000000000000116.

[7] Marczak ED, Marzec J, Zeldin DC, Kleeberger SR, Brown NJ, Pretorius M, Lee CR. Polymorphisms in the transcription factor NRF2 and forearm vasodilator responses in humans. Pharmacogenet Genomics. 2012;22(8):620–628. DOI:10.1097/FPC.0b013e32835516e5.

[8] Queisser N, Oteiza PI, Link S, Hey V, Stopper H, Schupp N. Aldosterone activates transcription factor Nrf2 in kidney cells both in vitro and in vivo. Antioxid Redox Sign. 2014;21(15): 2126–2142. DOI:10.1089/ars.2013.5565.

[9] Saw CL, Huang MT, Liu Y, Khor TO, Conney AH, Kong AN. Impact of Nrf2 on UVB-induced skin inflammation/photoprotection and photoprotective effect of sulforaphane. Mol Carcinog. 2011;50(6):479–486. DOI:10.1002/mc.20725

[10] Nojiri H, Saita Y, Morikawa D, Kobayashi K, Tsuda C, Miyazaki T, Saito M, Marumo K, Yonezawa I, Kaneko K, Shirasawa T, Shimizu T. Cytoplasmic superoxide causes bone fragility owing to low-turnover osteoporosis and impaired collagen cross-linking. J Bone Miner Res. 2011;26(11):2682–2694. DOI:10.1002/jbmr.489.

[11] Lobo RA, Davis SR, De Villiers TJ, Gompel A, Henderson VW, Hodis HN, Lumsden MA, Mack WJ, Shapiro S, Baber RJ. Prevention of diseases after menopause. Climacteric. 2014;17:540–556. DOI:10.3109/13697137.2014.933411.

[12] Clark NG, Fox KM, Grandy S; SHIELD Study Group. Symptoms of diabetes and their association with the risk and presence of diabetes: findings from the Study to Help Improve Early evaluation and management of risk factors Leading to Diabetes (SHIELD). Diabetes Care. 2007;30:2868–2873. DOI:10.2337/dc07-0816.

[13] Stachowiak G, Pertyński T, Pertyńska-Marczewska M. Metabolic disorders in menopause. Prz Menopauzalny. 2015;14(1):59–64. DOI:10.5114/pm.2015.50000.

[14] Derby CA1, Crawford SL, Pasternak RC, Sowers M, Sternfeld B, Matthews KA. Lipid changes during the menopause transition in relation to age and weight: the study of women's health across the nation. Am J Epidemiol. 2009;169:1352–1361. DOI: 10.1093/aje/kwp043.

[15] Matthews KA, Crawford SL, Chae CU, Everson-Rose SA, Sowers MF, Sternfeld B, Sutton-Tyrrell K. Are changes in cardiovascular disease risk factors in midlife women due to chronological aging or to the menopausal transition? J Am Coll Cardiol. 2009;54:2366–2373. DOI:10.1016/j.jacc.2009.10.009.

[16] Kabir F, Jahan N, Sultana N, Akter R. Lipid profile status in surgical menopause. J. Bangladesh Soc Physiol. 2011;6(2):127–133. DOI:10.3329/jbsp.v6i2.9763.

[17] Rossi R, Origliani G, Modena MG. Transdermal 17-beta-estradiol and risk of developing type 2 diabetes in population of healthy, nonobese postmenopausal women. Diabetes Care. 2004;27:645–649.

[18] Toth MJ, Sites CK, Poehlman ET, Tchernof A. Effect of menopausal status on lipolysis: comparison of plasma glycerol levels in middle-aged, premenopausal and early, postmenopausal women. Metabolism. 2002;51:322–326.

[19] Lambrinoudaki I. Cardiovascular risk in postmenopausal women with the polycystic ovary syndrome. Maturitas. 2011;68:13–16. DOI:10.1016/j.maturitas.2010.09.005.

[20] McGowan MP. Polycystic ovary syndrome: a common endocrine disorder and risk factor for vascular disease. Curr Treat Options Cardiovasc Med. 2011;13:289–301. DOI: 10.1007/s11936-011-0130-0.

[21] Schmidt J, Brännström M, Landin-Wilhelmsen K, Dahlgren E. Reproductive hormone levels and anthropometry in postmenopausal women with polycystic ovary syndrome (PCOS): a 21-year follow-up study of women diagnosed with PCOS around 50 years ago and their age-matched controls. J Clin Endocrinol Metab. 2011;96:2178–2185. DOI: 10.1210/jc.2010-2959.

[22] Schmidt J, Landin-Wilhelmsen K, Brännström M, Dahlgren E. Cardiovascular disease and risk factors in PCOS women of postmenopausal age: a 21-year controlled follow-up study. J Clin Endocrinol Metab. 2011;96:3794–3803. DOI:10.1210/jc.2011-1677.

[23] Waki H, Tontonoz P. Endocrine functions of adipose tissue. Annu Rev Pathol. 2007;2:31–56. DOI:10.1146/annurev.pathol.2.010506.091859.

[24] Otunctemur A, Dursun M, Ozbek E, Sahin S, Besiroglu H, Koklu I, Polat EC, Erkoc M, Danis E, Bozkurt M. Effect of metabolic syndrome on sexual function in pre- and postmenopausal women. J Sex Marital Ther. 2014;13:1–10. DOI:10.1080/0092623X.2014.918068.

[25] Prokai L, Rivera-Portalatin NM, Prokai-Tatrai K. Quantitative structure-activity relationships predicting the antioxidant potency of 17β-Estradiol-related polycyclic phenols to inhibit lipid peroxidation. Int J Mol Sci. 2013;14:1443–1454. DOI:10.3390/ijms14011443. DOI:10.1016/j.redox.2013.05.003.

[26] Bellanti F, Matteo M, Rollo T, De Rosario F, Greco P, Vendemiale G, Serviddio G. Sex hormones modulate circulating antioxidant enzymes: impact of estrogen therapy. Redox Biol. 2013;1:340–346. DOI:10.1016/j.redox.2013.05.003.

[27] Zhang T1, Liang X, Shi L, Wang L, Chen J, Kang C, Zhu J, Mi M. Estrogen receptor and PI3K/Akt signaling pathway involvement in S-(−)equol-induced activation of Nrf2/ARE in endothelial cells. PLoS One. 2013;8(11):e079075. DOI:10.1371/journal.pone.0079075.

[28] Jehn M, Clark JM, Guallar E. Serum ferritin and risk of the metabolic syndrome in U.S. adults. Diabetes Care 2004;27:2422–2428.

[29] Canoy D, Wareham N, Luben R, Welch A, Bingham S, Day N, Khaw KT. Serum lipid concentration in relation to anthropometric indices of central and peripheral fat distribution in 20,021 British men and women: results from the EPIC-Norfolk population-based cohort study. Atherosclerosis. 2006;189:420–427. DOI:10.1016/j.atherosclerosis.2005.12.027.

[30] Crist BL, Alekel DL, Ritland LM, Hanson LN, Genschel U, Reddy MB. Association of oxidative stress, iron, and centralized fat mass in healthy postmenopausal women. J Women's Health. 2009;18(6):795–801. DOI:10.1089=jwh.2008.0988.

[31] Cervellati C1, Bonaccorsi G, Cremonini E, Bergamini CM, Patella A, Castaldini C, Ferrazzini S, Capatti A, Picarelli V, Pansini FS, Massari L. Bone mass density selectively correlates with serum markers of oxidative damage in post-menopausal women. Clin Chem Lab Med. 2013;51(2):333–338. DOI:10.1515/cclm-2012-0095.

[32] Kanzaki H, Shinohara F, Kajiya M, Kodama T. The Keap1/Nrf2 protein axis plays a role in osteoclast differentiation by regulating intracellular reactive oxygen species signaling. J Biol Chem. 2013;288(32):23009–23020. DOI:10.1074/jbc.M113.478545.

[33] Hyeon S, Lee H, Yang Y, Jeong W. Nrf2 deficiency induces oxidative stress and promotes RANKL-induced osteoclast differentiation. Free Radic Biol Med. 2013;65:789–799. DOI:10.1016/j.freeradbiomed.2013.08.005.

[34] Ibáñez L, Ferrándiz ML, Brines R, Guede D, Cuadrado A, Alcaraz MJ. Effects of Nrf2 eficiency on one microarchitecture in an experimental model of osteoporosis. Oxid Med Cell Longev. 2014;ID 726590, 9 p. DOI:10.1155/2014/726590.

[35] Bouligand J, Cabaret O, Canonico M, Verstuyft C, Dubert L, Becquemont L, Guiochon-Mantel A, Scarabin PY. Estrogen and Thromboembolism Risk (ESTHER) Study Group. Effect of NFE2L2 genetic polymorphism on the association between oral estrogen therapy and the risk of venous thromboembolism in postmenopausal women. Clin Pharmacol Ther. 2011;89:60–64. DOI:10.1038/clpt.2010.241.

[36] Hosoya T1, Maruyama A, Kang MI, Kawatani Y, Shibata T, Uchida K, Warabi E, Noguchi N, Itoh K, Yamamoto M. Differential responses of the Nrf2–Keap1 system to laminar and oscillatory shear stresses in endothelial cells. J Biol Chem. 2005;280:27244–27250. DOI:10.1074/jbc.M502551200.

[37] Kalinowski L, Dobrucki IT, Malinski T. Race-specific differences in endothelial function: predisposition of African Americans to vascular diseases. Circulation. 2004;109:2511–2517. DOI:10.1161/01.CIR.0000129087.81352.7A.

[38] Bagat'e K, Meiring JJ, Cassee FR, Borm PJA. The effect of particulate matter on resistance and conductance vessels in the rat. Inhal Toxicol. 2004;16(6–7):431–436. DOI: 10.1080/08958370490439588.

[39] Terzano C, Di Stefano F, Conti V, Graziani E, Petroianni A. Air pollution ultrafine particles: toxicity beyond the lung. Eur Rev Med Pharmacol Sci. 2010;14(10):809–821.

[40] Mills NL, Donaldson K, Hadoke PW, Boon NA, MacNee W, Cassee FR, Sandström T, Blomberg A, Newby DE. Adverse cardiovascular effects of air pollution. Nat Clin Pract Cardiovascu Med. 2009;6(1):36–44. DOI:10.1038/ncpcardio1399.

[41] Zuo L, Youtz DJ, Wold LE. Particulate matter exposure exacerbates high glucose-induced cardiomyocyte dysfunction through ROS generation. PLoS ONE. 2011;6(8):e23116 DOI:10.1371/journal.pone.0023116.

[42] Howden R, Gougian E, Lawrence M, Cividanes S, Gladwell W, Miller-DeGraff L, Myers PH, Rouse DC, Devlin RB, Cho HY, Kleeberger SR. The Influence of Nrf2 on cardiac responses to environmental stressors. Oxid Med Cell Longev. 2013. ID 901239, 10 p. DOI:10.1155/2013/901239.

[43] Park JG, Oh GT. The role of peroxidases in the pathogenesis of atherosclerosis. BMB Rep. 2011;44(8):497–505.

[44] Sawyer DB. Oxidative stress in heart failure: what are we missing? Am J Med Sci. 2011;342(2):120–124. DOI:10.1097/MAJ.0b013e3182249fcd.

[45] Afanas'ev I. ROS and RNS signaling in heart disorders: could antioxidant treatment be successful? Oxid Med Cell Longev; 2011. ID 293769, 13 p. DOI:10.1155/2011/293769.

[46] Nigro P1, Abe J, Berk BC. Antioxid Redox Signal. 2011;15(5):1405–1414. DOI:10.1089/ars.2010.3679.

[47] Bailey-Downs LC, Mitschelen M, Sosnowska D, Toth P, Pinto JT, Ballabh P, Valcarcel-Ares MN, Farley J, Koller A, Henthorn JC, Bass C, Sonntag WE, Ungvari Z, Csiszar A. Liverspecific knockdown of IGF-1 decreases vascular oxidative stress resistance by impairing the Nrf2-dependent antioxidant response: a novel model of vascular aging. J Gerontol A Biol Sci Med Sci. 2012;67(4):313–329. DOI:10.1093/gerona/glr164.

[48] Konstantino Y, Nguyen TT, Wolk R, Aiello RJ, Terra SG, Fryburg DA. Potential implications of matrix metalloproteinase-9 in assessment and treatment of coronary artery disease. Biomarkers. 2009;14(2):118–129. DOI:10.1080/13547500902765140.

[49] Wu ML1, Ho YC, Yet SF. A central role of heme oxygenase-1 in cardiovascular protection. Antioxid Redox Signal. 2011;15(7):1835–1846. DOI:10.1089/ars.2010.3726.

[50] Howden R. Nrf2 and cardiovascular defense. Oxid Med Cel Longev. 2013, ID:104308, 10p. DOI:10.1155/2013/104308.

[51] Barajas B, Che N, Yin F, Rowshanrad A, Orozco LD, Gong KW, Wang X, Castellani LW, Reue K, Lusis AJ, Araujo JA. NF-E2-related factor 2 promotes atherosclerosis by effects on plasma lipoproteins and cholesterol transport that overshadow antioxidant protection. Arterioscler Thromb Vasc Biol. 2011;31(1):58–66. DOI:10.1161/ATVBAHA.110.210906.

[52] de Groot H, Rauen U. Ischemia-reperfusion injury: processes in pathogenetic networks: a review. Transplant Proc. 2007;39(2):481–484. DOI:10.1016/j.transproceed.2006.12.012.

[53] Calvert JW, Jha S, Gundewar S, Elrod JW, Ramachandran A, Pattillo CB, Kevil CG, Lefer DJ. Hydrogen sulfide mediates cardioprotection through Nrf2 signaling. Circ Res. 2009;105(4):365–374. DOI:10.1161/CIRCRESAHA.109.199919.

[54] Brewer AC, Murray TV, Arno M, Zhang M, Anilkumar NP, Mann GE, Shah AM. Nox4 regulates Nrf2 and glutathione redox in cardiomyocytes in vivo. Free Radic Biol Med. 2001;51(1):205–215. DOI:10.1016/j.freeradbiomed.2011.04.022.

[55] Rajasekaran NS, Varadharaj S, Khanderao GD, Davidson CJ, Kannan S, Firpo MA, Zweier JL, Benjamin IJ. Sustained activation of nuclear erythroid 2-related factor 2/antioxidant response element signaling promotes reductive stress in the human mutant protein aggregation cardiomyopathy in mice. Antioxid Redox Signal. 2011;14(6):957–971. DOI:10.1089/ars.2010.3587.

[56] Ungvari Z, Bailey-Downs L, Gautam T, Sosnowska D, Wang M, Monticone RE, Telljohann R, Pinto JT, de Cabo R, Sonntag WE, Lakatta EG, Csiszar A. Age-associated vascular oxidative stress, Nrf2 dysfunction, and NF-kappaB activation in the nonhuman primate Macaca mulatta. J Gerontol A Biol Sci Med Sci. 2011;66(8):866–875. DOI:10.1093/gerona/glr092.

[57] Ungvari Z, Bailey-Downs L, Sosnowska D, Gautam T, Koncz P, Losonczy G, Ballabh P, de Cabo R, Sonntag WE, Csiszar A. Vascular oxidative stress in aging: a homeostatic

failure due to dysregulation of NRF2-mediated antioxidant response. Am J Physiol; 2011;301(2):H363–H372. DOI:10.1152/ajpheart.01134.2010.

[58] Thornton MJ. Estrogens and aging skin. Dermatoendocrinol. 2013;5(2):264–270. DOI: 10.4161/derm.23872

[59] Reisman SA, Lee CY, Meyer CJ, Proksch JW, Ward KW Topical application of the synthetic triterpenoid RTA 408 activates Nrf2 and induces cytoprotective genes in rat skin. Arch Dermatol Res. 2014;306:447–454. DOI:10.1007/s00403-013-1433-7.

[60] Novo E, Parola M. Redox mechanisms in hepatic chronic wound healing and fibrogenesis. Fibrogenes Tissue Repair. 2008;1:1–58. DOI:10.1186/1755-1536-1-5.

[61] Lee Y, Shin JM, Jang S, Choi DK, Seo MS, Kim HR, Sohn KC, Im M, Seo YJ, Lee JH, Kim CD. Role of nuclear factor E2-related factor 2 (Nrf2) in epidermal differentiation. Arch Dermatol Res. 2014;306:677–682. DOI:10.1007/s00403-014-1470-x.

[62] Kokot A, Metze D, Mouchet N, Galibert MD, Schiller M, Luger TA, Böhm M. a-Melanocyte-stimulating hormone counteracts the suppressive effect of UVB on Nrf2 and Nrf-dependent gene expression in human skin. Endocrinology. 2009; 150:3197–3206. DOI:10.1210/en.2008-1315. Mol Carcinog.

[63] Tang L, Li J, Lin X, Wu W, Kang K, Fu W. Oxidation levels differentially impact melanocytes: low versus high concentration of hydrogen peroxide promotes melanin synthesis and melanosome transfer. Dermatology. 2012;224:145–153. DOI: 10.1159/000336777.

[64] Schafer M, Dütsch S, auf dem Keller U, Navid F, Schwarz A, Johnson DA, Johnson JA, Werner S. Nrf2 establishes a glutathione-mediated gradient of UVB cytoprotection in the epidermis. Genes Dev. 2010;24:1045–1058. DOI:10.1101/gad.568810.

[65] Jian Z, Li K, Liu L, Zhang Y, Zhou Z, Li C, Gao T. Heme oxygenase-1 protects human melanocytes from H2O2-induced oxidative stress via the Nrf2-ARE pathway. J Invest Dermatol. 2011;131:1420–1427. DOI:10.1038/jid.2011.56.

[66] Jian Z, Li K, Song P, Zhu G, Zhu L, Cui T, Liu B, Tang L, Wang X, Wang G, Gao T, Li C. Impaired activation of Nrf2-ARE signaling pathway undermines H2O2-induced oxidative stress response: a possible mechanism for melanocyte degeneration in Vitiligo. J Invest Dermatol. 2014;134:2221–2230. DOI:10.1038/jid.2014.152.

[67] Wolnicka-Glubisz A, Nogal K, Zadlo A, Plonka PM. Curcumin does not switch melanin synthesis towards pheomelanin in B16F10 cells. Arch Dermatol Res. 2015;307:89–98. DOI:10.1007/s00403-014-1523-1.

[68] Xiao L, Du Y, Shen, He Y, Zhao H, Li Z. TGF-beta 1 induced fibroblast proliferation is mediated by the FGF-2/ERK pathway. Front Biosci Landmark Ed. 2012;17:2667–2674.

[69] Patterson AD, Carlson BA, Li F, Bonzo JA, Yoo MH, Krausz KW, Conrad M, Chen C, Gonzalez FJ, Hatfield DL. Disruption of thioredoxin reductase 1 protects mice from

acute acetaminophen-induced hepatotoxicity through enhanced activity. Chem Res Toxicol. 2013;26:1088–1096. DOI:10.1021/tx4001013.

[70] Perry CD, Alekel DL, Ritland LM, Bhupathiraju SN, Stewart JW, Hanson LN, Matvienko OA, Kohut ML, Reddy MB, Van Loan MD, Genschel U. Centrally located body fat is related to inflammatory markers in healthy postmenopausal women. Menopause. 2008;15(4Pt 1):619–627. DOI:10.1097/gme.0b013e318159f1a2.

[71] Meakin PJ, Chowdhry S, Sharma RS, Ashford FB, Walsh SV, McCrimmon RJ, Dinkova-Kostova AT, Dillon JF, Hayes JD, Ashford ML. Susceptibility of Nrf2-null mice to steatohepatitis and cirrhosis upon consumption of a high-fat diet Is associated with oxidative stress, perturbation of the unfolded protein response, and disturbance in the expression of metabolic enzymes but not with insulin resistance. Mol Cell Biol. 2014;34(17):3305–3320. DOI:10.1128/MCB.00677-14.

[72] Kobayashi EH, Suzuki T, Funayama R, Nagashima T, Hayashi M, Sekine H, Tanaka N, Moriguchi T, Motohashi H, Nakayama K, Yamamoto M. Nrf2 suppresses macrophage inflammatoryresponse by blocking proinflammatory cytokine transcription. Nat Commun. 2016;7:11624. DOI:10.1038/ncomms11624.

[73] Eggler AL1, Savinov SN2. Chemical and biological mechanisms of phytochemical activation of Nrf2 and importance in disease prevention. Recent Adv Phytochem. 2013;43:121–155. DOI:10.1007/978-3-319-00581-2_7.

[74] Senger DR, Li D, Jaminet SC, Cao S. Activation of the Nrf2 cell defense pathway by ancient foods: disease prevention by important molecules and microbes lost from the modern western diet. PLoS One. 2016;11(2):e0148042. DOI:10.1371/journal.pone.0148042.

[75] de Ferrars RM, Cassidy A, Curtis P, Kay CD. Phenolic metabolites of anthocyanins following a dietary intervention study in post-menopausal women. Mol Nutr Food Res. 2014,58:490–502. DOI:10.1002/mnfr.201300322.

Chapter 7

The Effect of Nrf2 on Diabetic Complications

Angela Adamski da Silva Reis,
Rodrigo da Silva Santos, Aline Helena da Silva Cruz,
Elisangela Gomes da Silva,
Aparecido Divino da Cruz and
Gustavo Rodrigues Pedrino

Additional information is available at the end of the chapter

http://dx.doi.org/10.5772/66132

Abstract

The Nrf2 has been identified as a key molecular player in orchestrating adaptive cellular interactions following a wide spectrum of cellular conditions that could be either extracellular or intracellular. The encoded transcription factor regulates genes, which contain antioxidant response elements (ARE) in their promoters; many of these genes encode proteins involved in response to environmental stress, detoxifying enzymes, metabolic enzymes, injury, and inflammation, which includes the production of free radicals. The association between oxidative stress and inflammation with progression of diabetic nephropathy and cardiomyopathy has been described. The prevention of diabetic nephropathy and cardiomyopathy has become a global concern for those who are working in diabetic care management. Therefore, activation of Nrf2 has the potential to protect against macromolecular damage. Studies have demonstrated the beneficial role of Nrf2 induction in the prevention of DN. Upon exposure of cells to oxidative stress or electrophilic compounds, Nrf2 dissociates from Keap1 and translocates into the nucleus to bind to antioxidant-responsive elements in the genes encoding antioxidant enzymes. Upregulation of these Nrf2-dependent antioxidants promotes detoxification and anti-inflammatory function. Thus, the Nrf2 activators have been suggested for preventing diabetic nephropathy.

Keywords: Nrf2, diabetic complications, nephropathy diabetic

1. Introduction

The stimatives demonstraded that 10% of the global population will have diabetes by 2035 [1]. Hyperglycemia, hyperlipidemia, and inflammation are the metabolic abnormalities in diabetes and are involved with the development of reactive oxygen or nitrogen species (ROS or RNS) [2].

Diabetes is a strikingly heterogeneous disease with variable clinical, pathologic, and molecular features [1]. It has been described that high glucose induces the oxidative damage and it is related potentially on the diabetic complications in animal models [3].

Additionally, it is known that Nrf2 expression increases in response to oxidative stress [2]. The Nrf2 transcription factor (nuclear factor, erythroid 2derived 2-like 2 or nuclear factor erythroid 2-related factor 2) has been identified as a key molecular player in orchestrating adaptive cellular interactions following a wide spectrum of cellular conditions that could be either extracellular or intracellular [3].

Upregulation of Nrf2 and/or its downstream antioxidant genes in response to hyperglycemia is of growing interest in the clinical and research community [2]. The Nrf2 is a master regulator of cellular detoxification response and redox status and also provides a protective action from various oxidative stresses and damages [1, 2]. Given that we have yet to understand molecular mechanisms of diabetes complications, the results of studies may play an important role in elucidating the molecular pathways and get the development of strategies for prevention, treatment, and management of macrovascular and microvascular diseases (diabetes complications).

2. Implications of Nrf2 transcription factor in diabetic complications

The gene *Nrf2* also known as *NFE2L2* encodes a transcription factor that regulates genes that contain antioxidant response elements (ARE) in their promoters; these genes encode proteins involved in response to environmental stress, detoxifying enzymes, metabolic enzymes, injury, and inflammation, which includes the production of free radicals. During nonstressed conditions, the Nrf2 is inactive in the cytoplasm connected to protein Kelch-like ECH-associated protein 1 (Keap1), which prevents its translocation to the nucleus [4].

During oxidative stress, a signal that involves phosphorylation and/or redox modification is transduced to the Keap1/Nrf2 complex (see **Figure 1**), leading to its disruption and nuclear translocation of Nrf2. However, under basal conditions, Keap1 mediates rapid ubiquitination and subsequent degradation of Nrf2 by the 26S proteasome [3, 4]. Cullin 3-based ubiquitin E3 (Cul-E3) ligase complex ubiquitinates Nrf2, and Keap1 serves as a substrate adaptor, which facilitates the ubiquitination of Nrf2 by Cullin 3. As a result, Nrf2 has a short half-life that lasts only 20 min under normal conditions. Oxidative stress destroys critical cysteine residues in Keap1, disrupting the Keap1-Cul3 ubiquitination system. If Nrf2 is not ubiquitinated, it builds up in the cytoplasm and is translocated into the nucleus [4–6].

In the nucleus, Nrf2 combines with a small protein called Maf to form a heterodimer, and, by binding to the ARE in the upstream promoter region, it initiates the transcription of various cytoprotective genes including those encoding antioxidant and phase II detoxifying enzymes such as catalase (CAT), superoxide dismutases (SODs), heme oxygenase-1 (HO-1), NAD(P)H dehydrogenase (quinone) 1 (NQO-1), glutathione peroxidase-1 (GPx-1), glutathione S-transferase (GST), and γ-glutamylcysteine synthase (γ-GCS). The antioxidant response provided by the NFE2L2 and Keap1-NFE2L2/ARE signaling pathways protects the pulmonary, hepatic, digestive, neural and cardiovascular systems, and Nrf2 is considered a

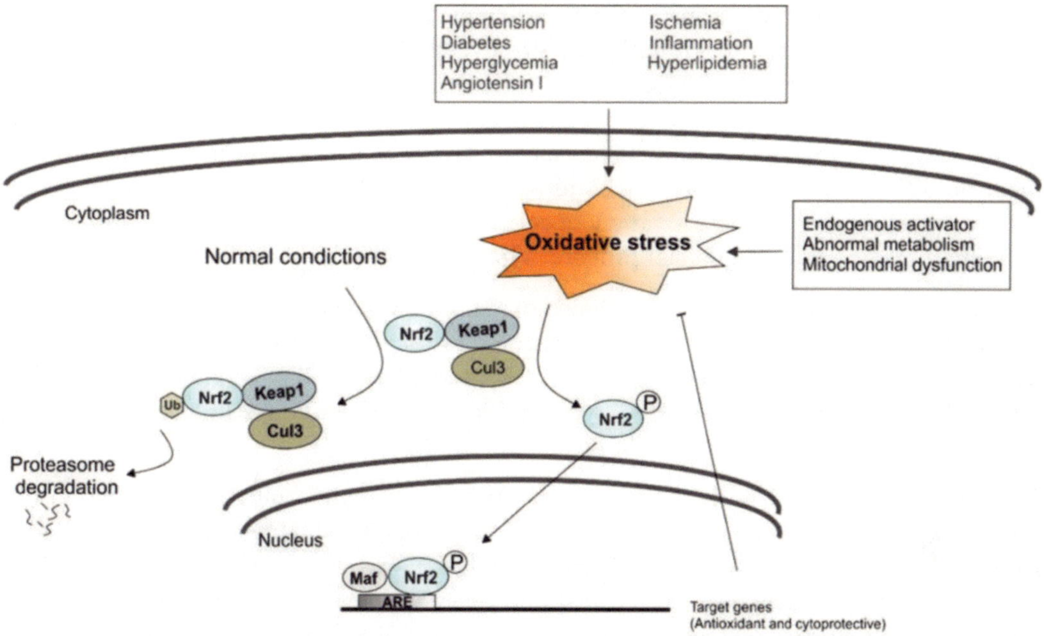

Figure 1. Schematic presentation of Nrf2 pathway activation by reactive oxygen species (ROS). Under basal conditions, Keap1 mediates rapid ubiquitination and subsequent degradation of Nrf2 by the 26S proteasome. During oxidative stress, the Nrf2 regulates genes, which contain antioxidant response elements (ARE) in their promoters.

promising target against diabetic complications such as cardiovascular diseases and diabetic nephropathy [7–9].

Type 2 diabetes mellitus (T2DM) is characterized by insulin resistance, with or without defects in insulin production and secretion, and as the prevalence of obesity in children has increased, T2DM has also become more common. T2DM is characterized by insulin resistance (with or without defects in insulin production and secretion), and the routine medical treatment is challenging, especially the identification and management of complications associated with micro- and macrovascular damage in diabetes [10].

Diabetes is a multifactorial process involving genetics, lifestyle, ethnic and racial heritage, and environmental factors. On the other hand, the interplay of these factors is not yet understood. The single nucleotide polymorphisms (SNPs) are a common type of genetic variations, which have been shown to impact most population susceptibility to diseases and individual response to drug treatments [10, 11].

In Chinese population was investigated NFE2L2 SNPs for possible associations with either T2DM or diabetic complications such as diabetic foot, microangiopathy and peripheral neuropathy, nephropathy, and retinopathy. This study showed that the T2DM patients with complications presented a higher frequency of mutant allele than the T2DM patients without complications. The study suggests that NFE2L2 SNPs are associated with T2DM patients with complications [10].

With oxidative stress and chronic systemic inflammation inseparably interconnected, inhibiting oxidative stress was theoretically an effective strategy to delay diabetes-related macrovascular and renal diseases. Research has revealed that the expressions of Nrf2-mediated anti-oxidative enzymes, including HO-1, NQO-1, and GPx-1, were significantly increased in the diabetic kidney when treated with salvianolic acid A (SAA) alone or in combination with metformin (MET). SAA is a polyphenol derivative extracted from the root of Salvia miltiorrhiza, which is known to show a variety of pharmacological activities including antioxidant, anti-inflammatory, and antiplatelet properties [8, 9].

Furthermore, SAA alleviates H_2O_2-mediated oxidative stress via activation of Nrf2/HO-1 signaling. The MET is an oral hypoglycemic agent, which is widely used in patients with type 2 diabetes. MET decreases blood glucose levels by decreasing hepatic glucose production, decreasing intestinal absorption of glucose, and improving insulin sensitivity by increasing peripheral glucose uptake and utilization [8–10].

Compared to treatment with either SAA or MET alone, their combination provided further protection against the macrovascular and renal injury, which was at least partly due to therapeutic activation of both MET-mediated AMP-activated protein kinase (AMPK) and SAA-mediated Nrf2/antioxidant response elements (ARE) pathways [12]. It is suggested that polyphenol Nrf2 modulators, especially combined with drugs activating AMPK including hypoglycemic drugs, are worthy of further investigation to combat diabetic complications [9].

The genetic deficiency in Nrf2-mediated transcriptional responses, especially antioxidant pathways, enhances susceptibility to both ischemic and nephrotoxic AKI in mice. Nrf2 deficiency is associated with an increased mortality accompanied by augmented kidney dysfunction and vascular permeability. Liu et al. [13] demonstrated that Nrf2 deficiency enhances susceptibility to both ischemic and nephrotoxic acute kidney injury and identifies this transcription factor as a potential therapeutic target in these injuries.

On the other hand, Nrf2 plays a protective role in experimental acute kidney injury, and this protection is mediated, in part, through an endogenous antioxidant pathway [2, 13].

Diabetic cardiomyopathy (DCM) is one of the major cardiac complications in diabetic patients. DCM is related to oxidative stress that is due to imbalance between ROS and/or reactive nitrogen species generation and their clearance by antioxidant defense systems [2–6]. Experimental and clinical studies have shown the important roles that Nrf2 and its downstream genes play in the pathogenesis of cardiac remodeling and heart failure induced by a number of factors. TNF-α has an important role in a number of pathologies associated with oxidative stress including diabetes, cancer, cardiac hypertrophy, and cardiomyopathy [6, 13].

The exposure of TNF-α to cells at concentrations well below the threshold associated with subinflammation significantly increased Nrf2 activity and its nuclear translocation. The TNFR1/2 double knockout mice and HL-1 cardiomyocytes make a first step forward in understanding the bimodal effects of the cytokine, TNF-α, in regulating the redox-sensitive Keap1/Nrf2 antioxidant pathway. This study has potential importance in the field of cardiovascular signaling because the TNF-α–induced biphasic regulation of the Nrf2 pathway suggests that a certain threshold of TNF/ROS signaling is essential to prime and activate the Nrf2 protective signaling pathway [14].

Research revealed that broccoli sprout extract (BSE), a natural SFN-rich supplement, can prevent the development of DCM. Like sulforaphane (SFN, an Nrf2 activator), BSE can prevent DCM in a transgenic T2DM mouse model via activation of Nrf2 by inhibiting diabetes-induced cardiac oxidative stress and damage as well as inflammation. BSE, when used at higher doses, can be used as a natural and safe source of SFN to upregulate Nrf2 expression and prevent DCM. Similar to its reported beneficial impact when used in patients with other chronic diseases, our results demonstrate that BSE also has promising potential in the treatment of patients with diabetes mellitus. Treatment with SFN-rich BSE for 3 months could significantly prevent the pathological process of DCM in the T2DM mice model, and like SFN, BSE also significantly upregulated Nrf2 expression and function to prevent diabetes-induced cardiac oxidative stress and inflammation. Therefore, BSE could potentially be used as a natural and safe treatment against DCM via Nrf2 activation [11].

Coronary artery disease and ischemic heart disease are prevalent worldwide. The development of percutaneous coronary intervention and surgical revascularization has brought marked benefits to patients with acute MI. However, ischemia/reperfusion injury during revascularization can cause further cardiac injury. Nrf2 and its target genes have been shown to play a protective role in cardiac ischemia-associated injury. Some antioxidants protect the heart from ischemia-induced cardiac injury via the Nrf2 pathway. For example, α-lipoic acid and prostaglandin D2 significantly increased Nrf2 nuclear translocation; the expression of its downstream genes reduced lactate dehydrogenase (LDH) and creatine kinase (CK) release, attenuated myocardial infarct size, decreased cardiomyocyte apoptosis, and partially preserved heart function; and this effect was at least partially PI3K/Akt signaling pathway dependent [6, 14, 15].

In adipose tissues, ROS promote the conversion from preadipocytes to mature adipocytes and facilitate insulin action [16, 17]. The ROS-mediated biological signaling pathways could be affected by enhanced Nrf2-ARE activity because ROS signaling intermediates should inversely correlate with the ROS scavenging activity and antioxidant status in cells. Thus, when the cells are chronically exposed to oxidative stressors, cellular ROS scavenging capacity is adaptively upregulated, primarily through the activation of Nrf2 and subsequent transcriptional induction of a suite of antioxidant enzymes [11].

Additionally, the induced antioxidant enzymes may have the undesired effect of impeding the physiological role of ROS as signaling molecules. Although antioxidants protect adipocytes from oxidative damage, they also may blunt aspects of ROS signaling, resulting in reduced adipogenesis and insulin resistance. Nrf2 controls white adipose tissue expandability and serves to maintain glucose and lipid homeostasis, including control of adipogenesis [19–21]. In addition, ROS, whose cellular concentrations decline with increases in Nrf2-regualted antioxidant gene expression, also affect insulin signaling [17, 18].

It is described that oxidative stress is a dynamic condition characterized by an imbalance between pro-oxidants and antioxidants. Reactive oxygen species (ROS), not adequately counterbalanced by antioxidant defenses, causes DNA damage. However, it is the action of antioxidants such as the Nrf2 transcription factor that acts by activating cytoprotective genes, which promote cell survival. In basal conditions, the repressor protein Keap1 binds Nrf2 in the cytoplasm and promotes its degradation. In the presence of ROS, Keap1 is inactivated and releases Nrf2 resulting in its nuclear translocation. The presence of mutations of the Nrf2-Keap1 genes and favoring greater Nrf2 expression and action have been described in association with different types of diseases. Recently, we have demonstrated the critical role of Nrf2 expression in protecting the cardiac cells from oxidative damage and death caused by high levels of glucose [19, 20, 22]. Thus, the cells have evolved endogenous defense mechanisms against sustained oxidative stress.

Experimental studies have shown that oxidative stress directly induces insulin resistance in cardiomyocytes via exaggerating extracellular signal-related kinase (ERK) activity in vitro. Additionally, depressed expression of cardiac Nrf2 was associated with significant increases in nitrosative damage and phosphorylation of ERK, all of which were prevented in the hearts of diabetic mice with cardiac overexpression of a potent antioxidant metallothionein (MT) [23, 24].

On the other hand, upregulation of cardiac Nrf2 by its activator dihydro-CDDO-trifluoroethyl amide (Dh404) significantly prevented diabetes-induced nitrosative damage, ERK activation, and insulin signaling downregulation. Thus, these findings suggest that oxidative stress–depressed expression of cardiac Nrf2 is associated with cardiac activation ERK and downregulation of glucose metabolism [25]. The studies suggested that the Nrf2 is a master transcriptional factor of antioxidative defense system [19] and may be a novel negative regulator of oxidative stress–mediated insulin resistance in cardiomyocytes and the heart [22].

Traditionally, interactions among metabolic and hemodynamic factors are considered to be involved in the development of renal lesions in patients with diabetes, for example, diabetic nephropathy (DN). However, several other factors such as oxidative stress and inflammatory processes have been shown to play important roles in the pathogenesis of DN; these factors are not completely independent, but they interact with each other. Several studies report the infiltration of macrophages and proinflammatory cells in kidney at different stages of DN [26]. The inflammatory infiltrate produces reactive oxygen species (ROS), proinflammatory cytokines, and growth factors, which lead to upregulation of chronic systemic inflammation and mediate the progression of diabetic nephropathy [26, 28–30].

As a consequence of inflammation, a variety of cytokines and acute phase proteins are released in order to augment or attenuate the inflammatory response. The main inflammatory cytokines involved in the development of DN are interleukin 1 (IL-1), IL-6, and IL-18 and tumor necrosis factor-α (TNF-α); these might contribute to the progression of renal injury either directly or indirectly [27]. Thus, chronic inflammation contributes to DN not only as a consequence of a direct effect of proinflammatory mediators on cellular signaling but also by creating a state of oxidative stress. In recent years, several investigators have provided substantial evidence implicating nuclear factor Nrf2 in inflammation and associated disorders [30, 31].

The therapeutic potential of Nrf2 activation in diabetes, implicating control of oxidative stress, in addition to regulation of inflammatory cytokines as methods of Nrf2 protection, was described. Also, it has been described that severe oxidative stress is associated with inflammation in chronic kidney disease (CKD) [27]. Oxidative stress and inflammation are mediators in the development and progression of chronic kidney disease (CKD) and its complications, and they are inseparably linked as each begets and amplifies the other. Although pathways involved in intrarenal ROS production and inflammation in experimental CKD have been widely explored, the relationship of DN on proinflammatory cytokines and Nrf2-Keap1 system in diabetes is poorly studied [27, 28, 30].

There was no an effective approach to prevent the development of these complications for the patients with diabetes. Thus, diabetic nephropathy and cardiomyopathy are the two major causes for the mortality of diabetic patients. Although there remain many questions to be further investigated, the potential beneficial effects of upregulation of Nrf2 and/or its downstream protective components have attracted the attention of basic researchers and clinical physicians to consider its potential application in the clinic [5–30].

3. Strategies to prevent diabetic complications by activating Nrf2 factor

The glucose control, blood pressure, lipid lowering, and the blockage of the renin-angiotensin systems were used for the treatment of diabetic patients, and the development and progression of nephropathy and cardiomyopathy in the patients with diabetes remain unpreventable [2]. Ultimately, it can be inferred that these Nrf2 cross talks with other signaling pathways are of clinical importance within the context of many human disease condition pathogenesis, particularly those with highly complex multifactorial molecular interaction, as in diabetic complications [32].

Diabetic nephropathy (DN), one of the major microangiopathic chronic diabetic complications, is associated with an increased risk of major cardiovascular events and all-cause mortality. DN is now the major cause of chronic kidney disease throughout the world and is the largest single cause of end-stage renal disease, accounting for nearly half of the patients entering dialysis each year. The etiopathogenesis of DN is clearly multifactorial, including genetic and environmental factors. Evidence indicates that mechanisms are active by mitochondrial overproduction of reactive oxygen species (ROS) [34, 37].

Extra generation of ROS, induced by hyperglycemia, is considered as the main reason for the development of these diabetic complications. Oxidative stress contributes to the pathogenesis of diabetic nephropathy. Nuclear factor erythroid-derived 2-like 2 (NRF2) controls cellular defense mechanisms against oxidative stress by turning on transcription of antioxidant genes [33]. Under physiological conditions Kelch-like ECH-associated protein 1 (Keap1) binds to Nrf2 and sequesters it in the cytoplasm. Under basal conditions, Keap1 mediates rapid ubiquitination and subsequent degradation of Nrf2 by the proteasome [7, 35, 36, 38]. Studies have demonstrated the beneficial role of Nrf2 induction in the prevention of DN [37].

Upon exposure of cells to oxidative stress or electrophilic compounds, Nrf2 dissociates from Keap1 and translocates into the nucleus to bind to antioxidant-responsive elements in the genes encoding antioxidant enzymes. Upregulation of these Nrf2-dependent antioxidants promotes detoxification and anti-inflammatory function. A growing body of evidence has indicated a critical role for activator-induced Nrf2 upregulation in the prevention of diabetic complications, including DN [36].

Other important targets of the Nrf2/Keap1/ARE system are lipogenic genes involved in regulation of triglyceride and cholesterol synthesis and metabolism, such as sterol response element-binding protein-1 and fatty acid synthase, where activation of Nrf2 leads to downregulation of lipogenic gene expression, affording protection against lipogenic stress. Repression of sterol response element-binding protein-1 decreased expression and activities of diacylglycerol acyltransferase-1 and 2 activity and fatty acid synthase, leading to decreased synthesis of triglycerides and cholesterol esters and secretion of apolipoprotein B100 in very low-density lipoprotein. Therefore, activation of Nrf2 has the potential to protect against macromolecular damage and against metabolic dysfunction and dyslipidemia [37].

Given these considerations, the preventive effect of Nrf2 activation on diabetic complication in animal models has been explored recently. The Nrf2 activators have been suggested for preventing diabetic nephropathy such as insulin, sulforaphane (SFN), cinnamic aldehyde (CA), and others. It is know that NQO1, one important Nrf2 downstream protective components, which is an important detoxifying enzyme that protective against diabetic complications. Functional variants of NQO1 were associated with the development of coronary artery disease in people with type 2 diabetes [35].

Studies have shown that after chronic treatment with SFN, that diabetic mice exhibited significant renal prevention from diabetes-induced damage most likely via induction of Nrf2-mediated antioxidant pathway. SFN has garnered particular interest as an indirect antioxidant due to its extraordinary ability to induce expression of several enzymes via the Keap1/Nrf2 pathway. The studies indicate the requirement of Nfr2 for SFN and CA-induced renal protection against diabetes [37]. The activation of Nrf2 by SFN is able to suppress hyperglycemia-induced oxidative stress and metabolic dysfunction in human microvascular endothelial cell [37].

Anti-inflammatory, antioxidant, and antimicrobial effects of curcumin products have been widely investigated recently [9–41]. The analog C66 from curcumin was found to effectively inhibit high glucose (HG)-induced inflammatory response and macrophage infiltration, resulting in a significant prevention of renal injury in diabetic rats, response, and macrophage infiltration, resulting in a significant prevention of renal injury in diabetic rats. Liu et al. [39] demonstrated for the first time that C66 can protect the aorta from diabetes using a type 1 diabetes mouse model. Treatment of diabetic mice with C66 for 3 months can almost completely reverse and/or prevent the progression of diabetes-induced aortic oxidative damage, inflammation, apoptosis, and proliferation.

C66 is a novel curcumin analog with a much lower effective dose of 5 mg/kg administered every other day and has been shown to establish protection from diabetic nephropathy and diabetic cardiomyopathy [7, 40]. Wu et al. [8] have demonstrated that in addition to upregulating Nrf2 by increasing miR-200a, C66 also protects against DN by inhibiting miR-21. Curcumin is a regulator of epigenetic events, including miRNAs, and among miR-21 has been demonstrated to play a key role in the pathogenesis of DN. Liu et al. described that C66's renal protection from diabetes was accompanied by a significant inhibition of c-Jun N-terminal kinase (JNK). Inhibition of JNK phosphorylation by C66 and JNKi also significantly prevented diabetes-induced increase in inflammation, oxidative and nitrative stress, apoptosis, cell proliferation, and fibrosis [39].

The health-beneficial effects of fruit and vegetables have been linked to their content of activators of Nrf2. Key dietary bioactive activators of this system are glucosinolate-derived dietary isothiocyanates (SFN) and indoles, thioethers and disulfides, polyphenols, flavonoids, carotenoids, oxidized omega-3 fatty acids, and triterpenoids. Additionally, SFN is found in the brassica vegetables (broccoli, cabbage, cauliflower, Brussel sprouts, and others) and rocket salad [11]. The diet as a source of anti-inflammatory and health-beneficial compounds for patients with chronic renal failure has been recognized. With markedly decreased clearance in this patient group and adverse effects of some dietary bioactive compounds at high doses, particularly glucosinolate-derived isothiocyanates, it is necessary to proceed cautiously with dietary recommendations [38].

The activation is thought to occur by different mechanisms: SFN releases Nrf2 from Keap1 by modification of critical cysteine thiol residues; some polyphenols induce downregulation of Keap1 expression and/or induce mild oxidative/nitrosative stress; and J3-isoprostanes disrupt the Keap1-Cul3 complex, preventing Keap1-Nrf2 targeting to the proteasome. Many chemically diverse activators have already been identified, including the glutathione peroxidase-1, SN found in cruciferous vegetables, caffeic acid phenethyl ester from the bee product propolis, and CA (found in cinnamon bark) [37]. Many have shown promising actions relevant to diabetes complications [38]. Dietary habits may also be difficult to change, but processing of food products and beverages may also be modified to increase levels of Nrf2 activators in foodstuffs. It is generally accepted that a diet rich in fruits and vegetables helps stave off the development of cardiovascular disease [38].

Dietary and synthetic activators use as secondary prevention measure for DN should remain a top priority for health official campaigns [41]. To reach the public health goal of reducing DN prevalence, campaigns to engaging in diabetic complications prevention need to be addressed, including dietary education strategies.

4. Conclusions

Oxidative stress is a major player in the etiology of diabetic complications. Given these considerations, the Nrf2 factor is a master regulator of redox homeostasis and the cellular detoxification response. It has been demonstrated that natural compounds derived from plants, vegetables, and micronutrients can activate Nrf2 and, thus, promote antioxidant pathways to mitigate oxidative stress and hyperglycemic damage. Studies are needed to evaluate the potential effect of Nrf2 activators. Thus, these activators play an important role in stress oxidative, such as a therapeutic strategy in preventing the development of diabetic complications.

Author details

Angela Adamski da Silva Reis[1,*], Rodrigo da Silva Santos[2], Aline Helena da Silva Cruz[1], Elisangela Gomes da Silva[1], Aparecido Divino da Cruz[3] and Gustavo Rodrigues Pedrino[4]

*Address all correspondence to: angeladamski@gmail.com

1 Department of Biochemistry and Molecular Biology, Biological Sciences Institute, Federal University of Goiás, Goiânia, Brazil

2 Department of Natural Sciences, Special Academic Unit of Human Sciences, Federal University of Goiás, Goiás, Brazil

3 Department of Medicine and Biology, Pontifical Catholic University of Goiás, Goiânia, Brazil

4 Department of Physiology, Biological Sciences Institute, Federal University of Goiás, Goiânia, Brazil

References

[1] Ghosh P, Sahoo R, Vaidya A, Chorev M, Halperin JA. Role of complement and complement regulatory proteins in the complications of diabetes. Endocrine Reviews. 2015;36(3):272–288. doi:10.1210/er.2014-1099

[2] Bing L, Liu S, Miao L, Cai L. Prevention of diabetic complications by activation of Nrf2: diabetic cardiomyopathy and nephropathy. Experimental Diabetes Research. 2012; 2012: Article ID 216512. doi:10.1155/2012/216512

[3] Ayres D, Baron B, Hunter T. miRNA Influences in NRF2 pathway interactions within cancer models. Journal of Nucleic Acids. 2015; 2015: Article ID 143636, 6 p. doi:10.1155/2015/143636

[4] Kobayashi M, Li L, Iwamoto N, Nakajima-Takagi Y, Kaneko H, Nakayama Y, Eguchi M, Wada Y, Kumagai Y, and Yamamoto M. The antioxidant defense system Keap1-Nrf2 comprises a multiple sensing mechanism for responding to a wide range of chemical compounds. Molecular and Cellular Biology. 2009; 29: 493–502. doi:10.1128/MCB.01080-08

[5] Choi BH, Kang KS, and Kwak MK. Effect of redox modulating NRF2 activators on chronic kidney disease. Molecules. 2014; 19(8): 12727–12759. doi:10.3390/molecules190812727

[6] Zhou S, Sun W, Zhang Z, Zheng Y. The role of Nrf2-mediated pathway in cardiac remodeling and heart failure. Oxidative Medicine Cellular Longevity. 2014; 2014: Article ID 260429. doi:10.1155/2014/260429

[7] Yamamoto T, Suzuki T, Kobayashi A, et al., Physiological significance of reactive cysteine residues of Keap1 in determining Nrf2 activity. Molecular and Cellular Biology. 2008; 28(8): 2758–2770. doi:10.1128/MCB.01704-07

[8] Wu P, Yan Y, Ma LL, Hou BY, He YY, Zhang L, Niu ZR, Song JK, Pang XC, Yang XY, Du GH. Effects of the Nrf2 modulator salvianolic acid a alone or combined with metformin on diabetes-associated macrovascular and renal injury. The Journal of Biological Chemistry. 2016; 291: 22288–22301. doi:10.1074/jbc.M115.712703

[9] Cheng JT, Huang CC, Liu I M, Tzeng TF, and Chang CJ. Novel mechanism for plasma glucose-lowering action of metformin in streptozotocin-induced diabetic rats. Diabetes. 2006; 55: 819–825.

[10] Ripsin CM, Kang H, Urban RJ. Management of blood glucose in type 2 diabetes mellitus. Am Fam Physician. 2009; 79: 29–36. http://www.aafp.org/afp/2009/0101/p29.html

[11] Xu Z, Wang S, Ji H, Zhang Z, Chen J, Tan Y, Wintergerst K, Zheng Y, Sun J, Cai L. Broccoli sprout extract prevents diabetic cardiomyopathy via Nrf2 activation in db/db T2DM mice. Scientific Reports. 2016; 6: 30252. doi:10.1038/srep30252

[12] Zhang H, Liu Y, Jiang Q, Li K.R. Zhao YX, Cao C, and Yao J. Salvianolic acid A protects RPE cells against oxidative stress through activation of Nrf2/HO-1 signaling. Free Radical Biology and Medicine. 2014; 69: 219–228. doi:10.1016/j.freeradbiomed.2014.01.025

[13] Liu M, Grigoryev DN, Crow MT, Haas M, Yamamoto M, Reddy SP, Rabb H. Transcription factor Nrf2 is protective during ischemic and nephrotoxic acute kidney injury in mice. Kidney International. 2009; 76(3): 277–285. doi:10.1038/ki.2009.157

[14] Shanmugam G, Narasimhan M, Sakthivel R, Kumar RR, Davidson C, Palaniappan S, Claycomb WW, Hoidal JR, Darley-Usmar VM, Rajasekaran NS. A biphasic effect of TNF-α in regulation of the Keap1/Nrf2 pathway in cardiomyocytes. Redox Biology. 2016; 9: 77–89 doi:10.1016/j.redox.2016.06.004

[15] Deng C, Sun Z, Tong G et al., Alpha-lipoic acid reduces infarct size and preserves cardiac function in rat myocardial ischemia/reperfusion injury through activation of pi3k/akt/nrf2 pathway. PLoS One. 2013; 8(3). Article ID e58371. doi:10.1371/journal.pone.0058371

[16] Lee H, Lee YJ, Choi H, Ko EH, Kim JW. Reactive oxygen species facilitate adipocyte differentiation by accelerating mitotic clonal expansion. Journal of Biological Chemistry. 2009; 284: 10601–10609. Epub 2009 Feb 23.

[17] Goldstein BJ, Mahadev K, Wu X, Zhu L, Motoshima H. Role of insulin-induced reactive oxygen species in the insulin signaling pathway. Antioxidants & Redox Signaling. 2005; 7: 1021–1031. doi:10.1089/ars.2005.7.1021

[18] Xue P, Hou Y, Chen Y, Yang B, Fu J, Zheng H, Yarborough K, Woods CG, Liu D, Yamamoto M, Zhang Q, Andersen ME, Pi J. Adipose deficiency of Nrf2 in ob/ob mice results in severe metabolic syndrome. Diabetes. 2013; 62(3): 845–54. doi:10.2337/db12-0584

[19] Li J, Ichikawa T, Villacorta L, et al. Nrf2 protects against maladaptive cardiac responses to hemodynamic stress. Arteriosclerosis Thrombosis and Vascular Biology Journal. 2009; 29: 1843–1850. doi:10.1161/ATVBAHA.109.189480

[20] He X, Kan H, Cai L, Ma Q. Nrf2 is critical in defense against high glucose-induced oxidative damage in cardiomyocytes. Journal of Molecular and Cellular Cardiology. 2009; 46: 47–58. . Epub 2008 Nov 1.

[21] Cheng X, Siow RC, Mann GE. Impaired redox signaling and antioxidant gene expression in endothelial cells in diabetes: a role for mitochondria and the nuclear factor-E2-related factor 2-kelch-like ECH-associated protein 1 defense pathway. Antioxidants & Redox Signaling. 2011: 14(3) 469–487. doi: 10.1089/ars.2010.3283..

[22] Tan Yi, Tomonaga Ichikawa, Jinqing Li, Qiusheng Si, Huaitao Yang, Xiangbai Chen, Curtis S. Goldblatt, Colin J. Meyer, Xiaokun Li, Lu Cai, and Taixing Cui. Diabetic downregulation of Nrf2 activity via ERK contributes to oxidative stress-induced insulin resistance in cardiac cells *in vitro* and *in vivo*. Diabetes. 2011; 60: 625–633. doi:10.1089/ars.2010.3283

[23] Cai L, Wang J, Li Y, et al. Inhibition of superoxide generation and associated nitrosative damage is involved in metallothionein prevention of diabetic cardiomyopathy. Diabetes. 2005; 54: 1829–1837. doi:10.2337/diabetes.54.6.1829

[24] Wang Y, Feng W, Xue W, et al. Inactivation of GSK-3beta by metallothionein prevents diabetes-related changes in cardiac energy metabolism, inflammation, nitrosative damage, and remodeling. Diabetes. 2009; 58: 1391–1402. doi:10.2337/db08-1697

[25] Li J, Ichikawa T, Janicki JS, Cui T. Targeting the Nrf2 pathway against cardiovascular disease. Expert Opinion on Therapeutic Targets. 2009; 13: 785–794. doi:10.1517/14728220903025762

[26] Galkina E, Ley K. Leukocyte recruitment and vascular injury in diabetic nephropathy. Journal of the American Society of Nephrology. 2006; 17(2): 368–377. doi:10.1681/ASN.2005080859

[27] Navarro-González JF, Mora-Fernández C. The role of inflammatory cytokines in diabetic nephropathy. Journal of the American Society of Nephrology. 2008; 19(3): 433–442. doi:10.1681/ASN.2007091048

[28] Pergola PE, Raskin P, Toto R.D. et al. Bardoxolone methyl and kidney function in CKD with type 2 diabetes. The New England Journal of Medicine. 2011; 365(4): 327–336. doi:10.1056/NEJMoa1105351

[29] Kim HJ, Vazir ND. Contribution of impaired Nrf2-Keap1 pathway to oxidative stress and inflammation in chronic renal failure. The American Journal of Physiology—Renal Physiology. 2010; 298(3): F662–F671. doi:10.1152/ajprenal.00568.2012

[30] Arellano-Buendía AS, Tostado-González M, García-Arroyo F Enrique, Cristóbal-García M, Loredo-Mendoza ML, Tapia E, Sánchez-Lozada LG, Osorio-Alonso H. Anti-inflammatory therapy modulates Nrf2-Keap1 in kidney from rats with diabetes. Oxidative Medicine and Cellular Longevity. 2016; 2016: Article ID 4693801: 3. doi:10.1155/2016/4693801

[31] Battelli QML, Hubbs AF. Multiorgan autoimmune inflammation, enhanced lymphoproliferation, and impaired homeostasis of reactive oxygen species in mice lacking the antioxidant-activated transcription factorNrf2. American Journal of Pathology. 2006; 168(6): 1960–1974.

[32] Li B, Liu S, Miao L, Cai L. Prevention of diabetic complications by activation of Nrf2: diabetic cardiomyopathy and nephropathy. Experimental Diabetes Research. 2012; 2012: Article ID 216512, 7 p. doi:10.1155/2012/216512

[33] Orlewski J, Orlewaska E. Effects of genetics polymorphisms of glutathione S-transferase genes (GSTM1, GSTT1, GSTP1) on risk of diabetic nephropathy: a meta-analysis. Polskie Archiwum Medycyny Wewnetrznej. 2015; 125(9): 649658. http://pamw.pl/en/issue/article/26252359

[34] Li B, Cui W, Tan Y, et al. Zinc is essential for the transcription function of Nrf2 in human renal tubule cells in vitro and mouse kidney in vivo under the diabetic condition. Journal of Cellular and Molecular Medicine. 2014; 18: 895–906. http://onlinelibrary.wiley.com/DOI/10.1111/jcmm.12239/full

[35] Zheng H, Whitman AS, Wu W, et al. Therapeutic potential of Nrf2 activations in streptozotocin-induced diabetic nephropathy. Diabetes. 2011; 60: 3055–3066. doi:10.2337/db11-0807

[36] Cui W, Bai Y, Miao X, Luo P, Chen Q, Tan Y, Rane MJ, Miao Lining, Cai Lu. Prevention of diabetic nephropathy by sulforaphane: possible role of Nrf2 upregulation and activation. Oxidative Medicine and Cellular Longevity. 2012; 2012: Article ID 821936, 12 p. doi:10.1155/2012/821936

[37] Thornalley PJ, Rabbani N. Dietary and synthetic activators of the antistress gene response in treatment of renal disease. Journal of Renal Nutrition. 2012; 22(1): 195–202. doi:10.1053/j.jrn.2011.10.01

[38] Haan JB. Nrf2 activators as attractive therapeutics for diabetic nephropathy. Diabetes. 2015; 60: 2683–2684. doi:10.2337/db11-0807

[39] Liu Y, Wang Y, Miao X et al. Inhibition of JNK by compound C66 prevents pathological changes of the aorta in STZ-induced diabetes. Journal Cellular Molecular Medicine. 2014; 18: 1203–1202. doi:10.1155/2014/289264

[40] Cai L, Wang Y, Zhou G, et al. Attenuation by metallothionein of early cardiac cell death via suppression of mitochondrial oxidative stress results in a prevention of diabetic

cardiomyopathy. Journal of the American College of Cardiology. 2006; 48: 1688–1697. doi:10.1016/j.jacc.2006.07.022

[41] McMahon M, Itoh K, Yamamoto M, Hayes JD. Keap1-dependent proteasomal degradation of transcription factor Nrf2 contributes to the negative regulation of antioxidant response element-driven gene expression. The Journal of Biological Chemistry. 2003; 278(24): 21592–21600. doi:10.1074/jbc.M300931200

Chapter 8

Nrf2 Signaling: An Adaptive Response Pathway for Neurodegenerative Disorders

Annalisa Lo Gerfo, Lucia Petrozzi, Lucia Chico and Gabriele Siciliano

Additional information is available at the end of the chapter

http://dx.doi.org/10.5772/65300

Abstract

Oxidative damage contributes to pathogenesis in many neurodegenerative diseases. As the indicator and regulator of oxidative stress, the nuclear factor erythroid 2-related factor 2 (Nrf2)-antioxidant response element (ARE) pathway has been shown to have dynamic changes and examined for its neuroprotective role in many cases. Nrf2 is emerging as a regulatory protein in neuronal death, since it helps neuronal cells to meet with oxidative insults. In this chapter, we summarize the role of Nrf2 as a master regulator of oxidative stress. Furthermore, we treat some natural and chemical substances able to modulate the Nrf2 pathway and, therefore, their possible use in the neurodegenerative diseases therapeutic treatment.

Keywords: cell metabolism, oxidative damage, neurodegenerative diseases, neuroprotection, modulators of Nrf2/ARE pathway

1. Introduction

To maintain redox homeostasis is very important for the normal function of the brain. This mechanism is regulated by antioxidant system. With age, genetic, and environmental risk factors, this system becomes imbalanced and oxidative stress (OS) follows through increased levels of reactive oxygen and nitrogen species (ROS/RNS). The accumulation of oxidative damage induces modifications of lipids, proteins, and DNA/RNA, a common feature of many neurodegenerative diseases, such as Parkinson's disease (PD), Alzheimer's disease (AD), amyotrophic lateral sclerosis (ALS), and Huntington's disease (HD). Although it is difficult to impose one if oxidative stress is a cause or epiphenomenon of neuronal death, the nuclear factor erythroid 2-related factor 2 (Nrf2)-antioxidant response element (ARE) pathway is a primary sensor of oxidative stress and regulates the expression of several genes encoding antioxidant proteins and detoxifying. The Nrf2-ARE pathway activation, in animal models of neurodegeneration, has produced positive effects; these data support the need for studies aimed to develop drugs able to activate Nrf2-ARE pathway in the central and peripheral nervous system. This chapter sums up the role of oxidative damage in neurodegenerative disorders and the protective functions of the Nrf2-ARE pathway.

2. Cell metabolism and oxidative stress

Originally, the primordial eukaryotic cells were unable to use oxygen for metabolic purposes. More than one billion years ago, according to endosymbiosis theory, these eukaryotic cells were colonized by aerobic bacteria, which can change with the host cells, and intracellular organelles became those we now call mitochondria. This alliance has facilitated bacteria to the availability of metabolic substrates, now assigned to the host cell, and at the same time has made a new kind of metabolism, much more efficient, eukaryotes: aerobic or oxidative metabolism [1]. Mitochondria are cytoplasmic organelles ranging from 1 to 10 μ and are often described as "electrical control units of the cell" because they generate most cell supply of adenosine triphosphate (ATP), used precisely as a chemical energy source from cell. Mitochondria not only perform this function but are also involved in other processes, such as signaling, cell differentiation, death, and also in the cell cycle control and cell growth [2]. The number of mitochondria in a cell varies according to the type of tissue and the body; there are cells with a single mitochondrion and cells with many thousands of mitochondria; these organelles are able to move freely in the cytoplasm and tend to thicken in the points where there is a greater demand for energy. The mitochondria, by means of the mitochondrial respiratory chain (OXPHOS), and through the process of oxidative phosphorylation, fulfill the requirements of ATP and therefore of energy of the cell [3]. The respiratory chain consists of a series of electron carriers (complexes), most of which are integral proteins of the inner membrane, containing prosthetic groups associated to proteins able to accept and donate one or two electrons [3]. The electron carrier complexes are four types: complexes I, II, III, and IV, in which two mobile electron carriers are to be added: cytochrome c and coenzyme Q. The respiratory chain is a very efficient mechanism, but during the step of transporting electrons, it may happen that a small percentage of electrons may prematurely reduce oxygen, forming reactive oxygen species (ROS), which are potentially harmful and dangerous for the cell. ROS are ions or very small molecules that include oxygen ions, free radicals and peroxides, organic and inorganic; they are highly reactive due to the presence of unpaired electrons in the orbital outside and are formed as a natural byproduct of oxygen metabolism and play an important role in cell signaling. The main source of ROS in vivo is aerobic respiration precisely, although they are also produced by the fatty acids beta-oxidation, by the xenobiotic components metabolism, after the activation of phagocytosis by pathogens. During periods of environmental stress, the ROS levels can increase dramatically, causing significant damage to cell structures. This increase is identified with the term of oxidative stress (OS) [4]. OS is usually defined as the altered balance between the production of ROS and their removal by cellular antioxidant mechanisms, such us enzymatic scavengers and low-molecular-weight reductants. Mitochondria use most of available oxygen (85–90%) to produce ATP, but, at the same time, are the major producers of ROS, such as superoxide (O–2) and hydrogen peroxide (H_2O_2) principally originate by loss of electrons from OXPHOS during oxidative phosphorylation with the consequent incomplete reduction of molecular oxygen [5, 6]. Superoxide itself is not greatly dangerous; nevertheless, it can rapidly react with the mild oxidant nitric oxide (NO to generate peroxynitrite (ONOO–) [7, 8]. Similarly, H_2O_2 is a slight oxidant but bit by bit it decomposes to generate the hydroxyl radical (·OH). Both ONOO– and ·OH damage the function of biomolecules inside the cell. Particularly, ROS attack the backbone and the side chains of proteins determining protein misfolding and aggregation. In addition, they attack nucleic acids, leading to alteration of purine and pyrimidine bases. Moreover, ROS cause lipid peroxidation, producing highly dangerous molecules, such as malondialdehyde, 4-hydroxy-2-trans-nonenal (HNE), acrolein, and thiobarbituric acid reactive substances (TBARSs) [9]. Summarizing, OS causes several interdependent mechanisms leading to cell death. All the human body's cells are subjected to oxidative stress, but the neurons are particularly affected by oxidative damage of aerobic metabolism. This susceptibility can be attributed on the one hand to their high oxygen requirement and on the other hand to low expression of antioxidant proteins [10]. Strong production of ROS is associated with deleterious effects on neuronal cell, also exerting crucial roles in regulating specific signaling mechanisms. In particular, ROS are able to activate kinase cascade [11], to regulate the calcium mobilization and signaling [12, 13], to control the expression of antioxidant genes [14, 15], and, finally, the ROS seem to control the differentiation [16] and neurogenesis [17] in neural stem cell. OS is a critical gambler in several diseases, including age-dependent neurodegenerative disorders such as Parkinson's disease (PD) and amyotrophic lateral sclerosis (ALS). The involvement of OS in several neurodegenerative conditions has been demonstrated by the identification of pathological mutations in genes performing in antioxidant pathways as well as oxidative stress markers in patients' samples [18–20]. Nevertheless, in many cases it is not clear whether this kind of stress is a primary cause or downstream event associated with the progression of the neurodegeneration. Consequently, a better understanding of ROS involvement in the pathogenesis of neurodegenerative diseases can offer the possibility to identify new targets for neuroprotective therapies.

3. Nrf2/ARE pathway

Several lines of evidence in the literature suggest that the reactive chemical species and electrophilic substances can have an important role in inducing different causative mechanisms of various pathologies such as tumorigenesis, diseases affecting the cardiovascular system, central nervous system, and peripheral nervous system [21, 22]. The human body, in order to neutralize these toxic substances, has developed a plethora of defense mechanisms [23]. Between the several mechanisms, the Nrf2-ARE pathway is now considered the most regulator of cellular defense mechanisms against oxidative stress [24].

Nrf2 be up to the Cap'n'collar (Cnc) transcription factor family and is considered the leader of the antioxidant response since it regulates the expression of several defensive genes [25, 26]. Nrf2 is a very unstable protein, typically present in association with its negative regulator Kelch-like ECH-associated protein 1 (Keap1), which acts as a molecular sensor of cellular oxidative stress. Under basal condition, Keap1 restrains Nrf2 in the cytoplasm leading to its degradation. Particularly, Keap1 acts as a connection protein between Nrf2 and the Cul3-based E3-ubiquitin ligase complex, promoting Nrf2 ubiquitination and consequent degradation by the 26S proteasome [27, 28]. Activation of Nrf2 involves its cytosolic stabilization; specific cysteine residues (Cys 151, Cys 273, and Cys 288) have been identified as direct sensors for electrophiles and oxidants; chemical modifications in these sensor residues cause a conformational change that produce the dissociation of Nrf2 from Keap1. Nrf2, detached from his repressors, translocates to the nucleus and binds its partner, small Maf protein. The heterodimer Nrf2-sMAF ultimately binds antioxidant response element (ARE) sequences leading to the expression of cytoprotective genes thus allowing cell to efficiently cope with endogenous stress and exogenous toxicants [29]. Nrf2 also is able to modulate the transcription of genes involved in mitochondrial biogenesis [30] (**Figure 1**).

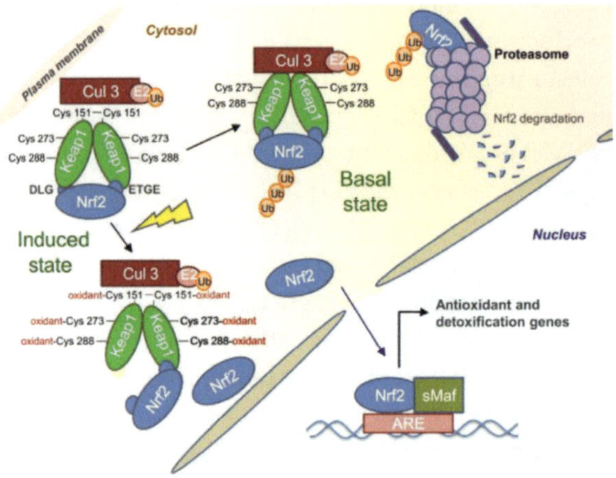

Figure 1. Regulation of Nrf2 by Keap1. In basal conditions, Nrf2 is sequestered in the cytoplasm by a Keap1 homodimer that facilitates ubiquitination and degradation of Nrf2 in the proteasome. In the presence of inducers that react with specific cysteine residues of Keap1, you get the release of Nrf2 and its nuclear translocation. In the nucleus, Nrf2 heterodimerizes with small Maf proteins and binds the antioxidant response element (ARE), by activating the expression of a battery of cytoprotective genes.

4. The Nrf2-ARE pathway and neuroprotection

4.1. Modifications of Nrf2-ARE pathway in neurodegenerative diseases

Abnormalities of Nrf2-ARE pathway were observed in several models of disease-aging dependent and in degenerative disorders; changes of Nrf2-ARE pathway cause ROS accumulation and therefore increase of oxidative damage to biological macromolecules. Several evidences in literature have showed that Nrf2 activation is induced in dopaminergic neurons in PD cases, but is decreased in hippocampus of AD patients [31]. In the motor cortex and spinal cord of ALS patients, *a* reduction of mRNA and protein levels of Nrf2 was observed; conversely, mRNA level of Keap1 is increased in the motor cortex [32]. Free radical scavengers regulated by Nrf2 such as superoxide dismutase 1 (SOD1) and catalase are reduced in patients, whereas other Nrf2-dependent genes are upregulated, for example, NAD(P)H:quinone oxidoreductase 1 (NQO1), an antioxidant enzyme, is upregulated in astrocytes, neurons, and other cell types in human AD [32–34] and PD brain [34]. Also heme oxygenase 1 (HO-1) is overexpressed in astrocytes and neurons of PD [35, 36] and AD patients [37]. HO-1 regulates heme degradation in two highly antioxidant molecules: biliverdin and bilirubin [38, 39]. Peroxiredoxin is able to reduce hydrogen peroxide and is also upregulated in PD [40], AD [41], and HD patients [42].

Pluripotent stem cells (iPSC)-derived neurons, which are generated from PD patients with PARK2 mutations, show an altered redox balance, mitochondrial dysfunction, and increased activity of the Nrf2 pathway [43]. Another study shows a significant alteration of the Nrf2-ARE pathway in neurospheres derived from the olfactory mucosa of PD patients [44]. Similar fluctuations are seen in two transgenic animal models of PD: αSyn-mutant A53T hαSynA53T [45] and MPTP mouse model [46]. In particular, Nrf2 downstream genes related to glutathione synthesis increase at the early stage and decrease at the late-stage disease with corresponding change in the total glutathione levels. Nrf2-regulated genes involved in glutathione synthesis, metabolism and transportation and detoxification of hydrogen peroxide/quinones increase in the SN and striatum of 1-month-old αSyn mice [47]. The Nrf2-ARE system is activated in cell culture systems as well in response to paraquat and maneb [48], and 6-hydroxydopamine [48, 49]. Apparently, the endogenous activation of the Nrf2-ARE pathway seems insufficient to neutralize the accumulation of oxidative damage. Thus, the research is focused in identifying an exogenous substance able to maintain long endogenous Nrf2-ARE activation to help the brain defend itself from oxidative damage [45, 50–53].

4.2. Nrf2-ARE pathway and Parkinson's disease

Parkinson's disease affects more than 1% of the population over 60 years of age and is the second most common neurodegenerative disorder after AD [54]. The 90% of cases are sporadic, whereas about 10% show a family background [55].

PD is caused by the degeneration of dopaminergic neurons within the substantia nigra pars compacta (SNc) and it is known that PD neurons are more susceptible to OS [56]. The selective vulnerability of SNc dopaminergic neurons can be caused by several pathogenic mechanisms that include exposure to genetic and environmental risk factors, altered proteolytic systems, and mitochondrial dysfunctions [57]. In particular, mitochondrial defects lead to impaired energy and ROS production and therefore to altered bioenergetic and redox balance.

Consistent evidence shows that disrupted mitochondrial integrity and OS play a pivotal role in PD pathogenesis and disease progression.

Mutations in several genes, such as PARK2, PARK6, and PARK7, are associated with early-onset familial forms of PD and with mitochondrial alterations leading to neuronal death [58]. Several studies show that some substances, such as MPTP and rotenone, are able to inhibit mitochondrial complex I and increase ROS production with possible loss of dopaminergic neurons in the SNc [59]. Furthermore, peripheral and central markers of oxidative damage are altered in PD patients, indicating that OS is a crucial player in PD pathogenesis [60–62]. PD has also been associated with alterations in the expression of antioxidant molecules such as glutathione and antioxidant enzymes. It was shown that oxidized glutathione is significantly higher, while other antioxidant molecules and catalase activity are decreased in blood cells from PD patients [63]. Furthermore, several studies have shown that the activation of antioxidant genes expression, in particular those under the control of the Nrf2/ARE system, has neuroprotective effects in different models of PD [64, 65].

Activation of the Nrf2-ARE pathway is able to protect against the toxic forms of αSyn. In SK-N-SH neuroblastoma cells, ferrous iron promotes αSyn aggregation through inhibiting Nrf2 pathway [66]. In a PD animal model, it was observed that the transgenic activation of Nrf2 and knockdown of Keap1 could delay the αSyn-mediated dopaminergic neuron loss and motor dysfunction [67]. Conversely, genetic deletion of Nrf2 increases αSyn toxicity and exaggerates αSyn/p-αSyn accumulation in dopaminergic neurites and gliosis. Nrf2 deficiency enhances the inflammatory response and lowers the capability of phagocytosis in primary microglial cells [68].

Recently, studies have identified the importance of astrocytic Nrf2-regulating αSyn proteostasis. Astrocytic overexpression of Nrf2 (GFAP-Nrf2) can reduce αSyn aggregates in the central nervous system of a PD mouse model with neuronal overexpression of human αSyn-mutant A53T [45]. The accumulation of hαSynA53T in the Triton-soluble fraction from the spinal cord decreases 60% in symptomatic mice. This is accompanied by a significant increase in hαSynA53T in the Triton-insoluble/SDS-soluble fraction. This movement of αSynA53T into Triton-insoluble/SDS-soluble aggregates is completely reversed by the overexpression of Nrf2 in astrocytes.

Similar changes are observed for phosphorylated (Ser129) αSynA53T (p-hαSynA53T) in Triton-soluble and Triton-insoluble/SDS-soluble fractions. Fluorescent staining of hαSynA53T also shows a dramatic increase in hαSynA53T aggregates that colocalized with phαSynA53T. Again, these changes are completely reversed by GFAP-Nrf2.

The autophagy-lysosome pathway (ALP) is a protein degradation system responsible for the turnover of proteins, aggregate proteins, and damaged organelles. Dysfunctions of autophagic mechanism result in the accumulation of cytoplasmic aggregates composed of misfolded proteins and deformed organelles, leading to neurodegeneration and other diseases [69–71]. A significant dysfunction of autophagic machinery is observed in the αSynA53T mice model [45, 72–74]. Nrf2 prevents chaperone-mediated autophagy dysfunction and increases lifespan, delays onset, and reduces aggregation in αSynA53T mice [45].

4.3. Nrf2-ARE pathway and Alzheimer's disease

Alzheimer's disease (AD) is the most common neurodegenerative disease, accounting for 60–70% of cases of dementia, and although its etiology is still unclear, it is characterized by the presence of brain amyloid plaques and neurofibrillary tangles whose accumulation ultimately leads to extensive neuronal loss and progressive decline of cognitive function [75–77]. They are aggregates of proteins distributed in the entorhinal cortex, hippocampus, and temporal, frontal, and inferior parietal lobes. Amyloid plaques are composed of aggregates of β-amyloid (Aβ) and of other protein aggregates such as hyperphosphorylated Tau, ubiquitin, and presenilins 1 and 2; amyloid plaques are sticky buildup which accumulates outside nerve cells. Neurofibrillary tangles are abnormal collections of twisted protein threads found inside nerve cells; the tangles are aggregates of hyperphosphorylated Tau protein [78]. Some of the major risk factors for AD are unhealthy aging in sporadic AD cases, the presence of ApoE-4 alleles in both sporadic and familial AD [79], and genetic factors, such as mutation in amyloid precursor protein (APP) and presenilin-1 (PS1) in familial AD [80] among others. AD brain is characterized by mitochondrial dysfunction, reactive gliosis, and oxidative damage to lipids and proteins [81–85].

Several studies demonstrate that the AD brain is under oxidative stress attack. A significantly increased HO-1 expression was reported in postmortem AD temporal cortex and hippocampus [82]. Additionally, an increased Nqo1 activity and expression were found in astrocytes and neurons of AD brain [86, 87] and Nrf2 was localized in cytoplasm in AD hippocampal neurons [31]. Furthermore, there is increased protein oxidation [88, 89] and lipid peroxidation [90–92] in AD brain. Recent studies in aged APP/PS1 AD mouse models showed Nrf2 protein levels [93].

Several evidences in literature have shown that Nrf2-ARE pathway is able to mitigate the toxicity mediated by Aβ. These studies have confirmed the neuroprotective role of Nrf2 against ROS generation and cell death induced by Aβ in vitro [94–97]. Tert-butylhydroquinone (tBHQ), a prototypical Nrf2 activator, has been reported to reduce Aβ1-42 secretion in the NT2N cell line with increased cell viability [98]. The sulforaphane, Nrf2 activator, is able to preserve cognitive function in an AD animal model [99]. Strikingly, overexpression of mitochondria catalase in APP (Tg2576) transgenic mice dramatically reduces full-length APP and its c-terminal fragment 99, lowers soluble and insoluble Aβ levels, extends lifespan, and improves working memory [100]. A genetic study has demonstrated that overexpression of Nrf2 in the hippocampus causes increase in mTOR activity; these data suggest that Nrf2 could mediate autophagy and alter processing/clearance of APP and/or Aβ.

4.4. Nrf2-ARE pathway and amyotrophic lateral sclerosis

ALS is a progressive disease with fatal outcome, in which the motor cortex and spinal cord motor neurons are selectively affected. The disease in 90% of cases occur sporadically (SALS) while in 10% of cases there is a clear familiarity (FALS) [101]. The etiology and pathogenesis of ALS are currently largely unknown. ALS is considered a degenerative multifactorial disease in which cell death is a consequence of a complex interaction between genetic risk factors and environmental factors. To explain the neuronal death, several hypotheses have been proposed, among which the most accredited implicates oxidative stress [102–106]. In fact, levels of oxidative stress biomarkers were observed to be altered in SALS patients; these data indicate that most likely a redox imbalance is relevant in the pathogenesis of disease [107–112]. Elevated levels of HNE have been detected also in cerebrospinal fluid (CSF) from ALS patients [113, 114]. Additionally, mitochondrial alterations have been observed in motor neuron of ALS patients [115–118]. These dysfunctions are tightly interrelated with OS cascades, activating overlapping molecular pathways in a vicious cycle of harmful events. Specifically, alterations in mitochondrial morphology and biochemistry have been extensively detected in postmortem tissues [119] and in lymphocytes [120] from SALS patients, in SOD1 transgenic mice, and cellular models [52]. Dynamic and morphological abnormalities, along with metabolic deficits in the activities of the OXPHOS proteins, have also been described in both SALS and FALS patients [121]. Furthermore, impairment in antioxidant mechanisms has also been shown in ALS, including downregulation of members of glutathione S-transferase family [122, 123], peroxiredoxins [124], and Nrf2 [125–129].

The first causative gene associated with genetic ALS form was the Cu-Zn superoxide dismutase 1. In FALS patients with SOD1 gene mutations and in G85R animal model, cytoplasmic inclusions containing modified SOD1 proteins have been observed [130]. In the last decade, genome-wide association (GWA) studies identified two genes associated with sporadic and non-SOD1 familial ALS: RNA/DNA-binding proteins, 43-kDa transactive response (TAR) DNA-binding protein (TDP-43), and fused in sarcoma/translocated in liposarcoma (FUS/TLS) [131–135]. Both TDP-43 and FUS are predominantly nuclear proteins involved in RNA metabolism; however, both are observed as aggregates in the cytosol of ALS neurons [136].

Nrf2 activators have been shown to protect against oxidative stress and cell death induced by SOD1-mutant protein [137, 138]. The Nrf2 overexpression in glial cells directly increases the resistance to oxidative stress and helps indirectly, through the increate secretion of glutathione, the ability of the motor neurons to neutralize the toxic effects caused by SOD1-mutant protein [139]. Also Nrf2 and Keap1 expression analysis showed a reduction of Nrf2 protein in patients than in controls; conversely, there have been no significant differences in the expression of Keap1 levels between patients and controls [140–142].

Recently, NSC34 motor neuronal cell lines expressing TDP-43 mutants exhibit shortened neurites, alteration of oxidative stress markers levels. These effects are reversed by the UPS inhibitor MG132, but not by the Nrf2 activator sulforaphane [143, 144]. This is attributed to an increase in HO-1 following MG132 treatment that appeared to be independent of Nrf2 activation. While the role of Nrf2 in protection against SOD1-mutant neuronal toxicity is clear, its effect on other ALS-associated gene mutations particularly TDP43 and FUS needs to be clarified by future studies.

5. Modulators of Nrf2/ARE pathway

The manipulation of the Nrf2-ARE pathway at the genetic level is being studied through the use of siRNA or antisense oligonucleotides against Keap1 to activate/overexpress Nrf2.

Antisense drugs are being researched to study neurodegenerative disorders, cancer, metabolic disorders, and disorders with inflammatory components among others. Antisense drug fomivirsen, marketed as Vitravene, has been approved by the US Food and Drug Administration (FDA) for the treatment of cytomegalovirus retinitis. Since then, numerous antisense therapies have been tested but have not produced significant clinical result. This has not diminished the potential of gene therapies. Antisense oligonucleotide can bind to the target RNA and disrupt RNA splicing, transcription, translation, and replication, thereby modulating gene expression. Several studies showed that siRNA-mediated knockdown of Keap1-activated Nrf2-ARE pathway in mouse cortical astrocytes and provided partial protection against MPTP-mediated toxicity in mouse, in vivo [65, 145]. The overexpression of target gene can also be achieved by viral-mediated gene transduction but it is too early to conclude on efficacy of viral-mediated gene therapy in human neurodegenerative disorder cases. Nrf2 modulation in various neurodegenerative disorders has been previously described in this chapter. Hence, using the antisense oligonucleotide against Keap1, lentiviral-mediated Nrf2 overexpression or siRNA against Keap1-mediated overexpression of Nrf2 treatment can prove beneficial in neurodegenerative disorders.

Among recent patents, Curna, Inc. filed patent for the use of antisense for the treatment of Nrf2-related disorders. The initial study published under International Application for the Patent Cooperation Treaty (PCT) showed that antisense CUR-0330 and CUR 0332 showed two- to threefold increase in Nrf2 mRNA expression compared to control (PCT/US2010/027394). The invention is targeted at the inhibition of natural antisense transcript to Nrf2 as a strategy toward modulation of Nrf2 expression in disease models [145].

The modulation of Nrf2 expression by using several other pharmacological interventions to inhibit Keap1 and Nrf2 interaction is under investigation.

The Nrf2/ARE pathway can be pharmacologically activated also by molecules of both natural derivation (nutraceuticals) and chemical synthesis. Between Nrf2/ARE activators of natural origin, sulforaphane, polyphenols, and curcumin have been included; between chemical synthesis substances, chemical Nrf2/ARE activators include triterpenoids and N-(4-(2-pyridyl)(1,3-thiazol-2-yl))-2-(2,4,6-trimethylphenoxy) acetamide (CPN-9).

SFN, derived from cruciferous vegetables such as broccoli, activates Nrf2 through the modification of reactive cysteine residues of Keap1 [146, 147], and SFN is able to overstep the blood-brain barrier, induce the transcription of Nrf2-dependent gene expression in the basal ganglia, and protect dopaminergic neurons from cell death MPTP induced [64, 148]. Other Nrf2/ARE pathway natural inducers are EGCG and resveratrol, belonging to the family of polyphenols that, for their antioxidant qualities, are considered to be important nutraceuticals. EGCG, a flavonoid polyphenol, for example, showed antioxidant and neuroprotective functions in cultured motoneuron-neuroblastoma hybrid cell line transfected with mutSOD1 [149] and in PC12 cells exposed to paraquat [150]. Furthermore, EGCG was shown to be neuroprotective in mice model of ALS: oral administration to mice expressing mutSOD1 delayed symptoms onset [151–154].

Resveratrol, a polyphenolic compound present in red wine, demonstrated protective effects against hypoxic injury in rat spinal cord dorsal column by activating Nrf2 pathway [155, 156].

Curcumin, a member of the curcuminoid family isolated from plant *Curcuma longa*, showed Nrf2-dependent antioxidant properties in primary spinal cord astrocytes exposed to H_2O_2 [157] and in ischemic brain injury models [158]. Other nutraceuticals, such as naphthazarin, genistein, and carnosic acid, showed positive effects in several models of neurodegenerative and cardiovascular diseases implicating OS as a pathogenic factor [148, 159–164].

Furthermore, several synthetic Nrf2/ARE activators were recently developed. Recently, triterpenoids emerged as a potent class of Nrf2/ARE inducers. Triterpenoids are very powerful inducer of Nrf2 pathway: they are able to protect dopaminergic neurodegeneration in MPTP mouse model of PD [165], and increase the lifespan in ALS mouse models [166]. Another chemical activator of Nrf2/ARE pathway is CPN-9 which selectively suppresses cell death triggered by OS in a cell-type-independent manner. SH-SY5Y cells pretreated with CPN-9 were more resistant to cytokine-induced apoptosis. CPN-9 is able to decrease the ROS levels through the induction of several antioxidant genes [137]. Finally, we know that that some drugs such as bromocriptine [167] and azathioprine [168] were capable to induce the Nrf2/ARE pathway, therefore providing insight into a possible development of new synthetic molecules Nrf2 activators.

6. Conclusion

Oxidative stress and misfolded proteins are two mechanisms that act together to the pathogenesis of several inflammatory and degenerative diseases. The detailed mechanism by which Nrf2-ARE pathway carries out its action is still unclear. Current data suggest that Nrf2 affects both primary protein degradation pathways, the UPS and ALP, which are both altered in neurodegenerative diseases.

Despite the progress made in understanding the importance of Nrf2/ARE pathway, it remains to clarify the exact mechanism by which it exerts its function so that it may lead to discovery of new targets for the treatment of neurodegenerative diseases. In the past decade, Nrf2-ARE pathway activation has shown promising results for the treatment of many disorders including neurodegenerative disease. Several of these Nrf2 activators or their brain accessible synthetically modified compounds have passed phase II and III clinical trials. BG-12, an oral formulation of DMF (Biogen Idec, Inc.), is in phase III clinical trials for the treatment of multiple sclerosis (MS). Bardoxolone methyl, an oral formulation of CDDO-MA (Reata Pharmaceuticals, Inc.), is currently in phase III clinical trials for chronic kidney disease in type II diabetes mellitus patients, but there are no existing clinical trials in the pipeline for neurodegenerative disorders. EGCG, resveratrol, and curcumin are in various phases of clinical trial for treatment and efficacy in neurodegenerative disorders such as AD, PD, and ALS. The knowledge gained from these studies will further help in identifying clinically relevant approaches for the activation of Nrf2 in CNS and potentially lead to finding treatments for these devastating neurological disorders.

http://dx.doi.org/10.5772/65300

Acknowledgements

This chapter was supported by the CariLucca Foundation grant (539999_2014_Siciliano_SLAF-CARILUCCA).

Conflict of Interest and Sources of Funding Statement: The authors declare that they do not have conflicts of interest.

Author details

Annalisa Lo Gerfo*, Lucia Petrozzi, Lucia Chico and Gabriele Siciliano

*Address all correspondence to: annalisalogerfo2@virgilio.it

Department of Clinical and Experimental Medicine, Neurological Clinic, University of Pisa, Pisa, Italy

References

[1] L. Ernster and G. Schatz, "*Mitochondria: a historical review*", J Cell Biol, 91: 227–255, 1981.

[2] G. T. Babcock and M. Wikstrom, "*Oxygen activation and the conservation of energy in cell respiration*", Nature, 356: 301–309, 1992.

[3] Y. Hatefi "*The mitochondrial electron transport and oxidative phosphorylation system*", Ann Rev Biochem, 54: 1015–1969, 1985.

[4] K. Maiese, Z. Z. Chong, J. Hou, Y. C. Shang, "*Erythropoietin and oxidative stress*", Curr Neurovasc Res, 5: 125–142, 2008.

[5] M. S. Hernandes and L. R. Britto, "*NADPH oxidase and neurodegeneration*", Curr Neuropharmacol, 10: 321–327, 2012.

[6] F. Orsini, M. Moroni, C. Contursi et al., "*Regulatory effects of the mitochondrial energetic status on mitochondrial p66Shc*", Biol Chem, 387: 1405–1410, 2006.

[7] W. A. Pryor and G. L. Squadrito, "*The chemistry of peroxynitrite: a product from the reaction of nitric oxide with superoxide*", Am J Physiol, 268: L699–L722, 1995.

[8] M. C. Martinez and R. Andriantsitohaina, "*Reactive nitrogen species: molecular mechanisms and potential significance in health and disease*", Antioxid Redox Signal, 11: 669–702, 2009.

[9] D. Trachootham, W. Lu, M. A. Ogasawara, R. D. Nilsa, P. Huang, "*Redox regulation of cell survival*", Antioxid Redox Signal, 10: 1343–1374, 2008.

[10] B. Halliwell, *"Role of free radicals in the neurodegenerative diseases: therapeutic implications for antioxidant treatment"*, Drugs Aging, 18: 685–716, 2001.

[11] Y. Son, Y. K. Cheong, N. H. Kim, H. T. Chung, D. G. Kang, H. O. Pae, *"Mitogen-activated protein kinases and reactive oxygen species: how can ROS activate MAPK pathways?"*, J Signal Transduct, 2011: 792639, 2011.

[12] Y. Yan, C. L. Wei, W. R. Zhang, H. P. Cheng, J. Liu, *"Crosstalk between calcium and reactive oxygen species signaling"*, Acta Pharmacol Sin, 27: 821–826, 2006.

[13] R. F. Feissner, J. Skalska, W. E. Gaum, S. S. Sheu, *"Crosstalk signaling between mitochondrial Ca2+ and ROS"*, Front Biosci, 14: 1197–1218, 2009.

[14] Q. Ma, *"Transcriptional responses to oxidative stress: pathological and toxicological implications"*, Pharmacol Ther, 125: 376–393, 2010.

[15] R. G. Allen and M. Tresini, *"Oxidative stress and gene regulation"*, Free Radical Biol Med, 28: 463–499, 2000.

[16] H. L. A. Vieira, P. M. Alves, A. Vercelli, *"Modulation of neuronal stem cell differentiation by hypoxia and reactive oxygen species"*, Prog Neurobiol, 93: 444–455, 2011.

[17] K. A. M. Kennedy, S. D. E. Sandiford, I. S. Skerjanc, S. S. C. Li, *"Reactive oxygen species and the neuronal fate"*, Cell Mol Life Sci, 69: 215–221, 2012.

[18] B. Halliwell, *"Oxidative stress and neurodegeneration: where are we now?"*, J Neurochem, 97: 1634–1658, 2006.

[19] A. Melo, L. Monteiro, R. M. F. Lima, D. M. de Oliveira, M. D. de Cerqueira, R. S. El-Bachà, *"Oxidative stress in neurodegenerative diseases: mechanisms and therapeutic perspectives"*, Oxid Med Cell Longev, 2011: 467180, 2011.

[20] S. Gandhi and A. Y. Abramov, *"Mechanism of oxidative stress in neurodegeneration"*, Oxid Med Cell Longev, 2012: 428010, 2012.

[21] R. Kohen and A. Nyska, *"Oxidation of biological systems: oxidative stress phenomena, antioxidants, redox reactions, and methods for their quantification"*, Toxicol Pathol, 30: 620–650, 2002.

[22] T. Nguyen, P. J. Sherratt, H.C. Huang, C. S. Yang, C. B. Pickett, *"Increased protein stability as a mechanism that enhances Nrf2-mediated transcriptional activation of the antioxidant response element. Degradation of Nrf2 by the 26 S proteasome"*, J Biol Chem, 278: 4536–4541, 2003.

[23] G. P. Sykiotis and D. Bohmann, *"Stress-activated cap'n'collar transcription factors in aging and human disease"*, Sci Signal, 3(112): re3, 2010.

[24] D. A. Johnson and J. A. Johnson, *"Nrf2-a therapeutic target for the treatment of neurodegenerative diseases"*, Free Radic Biol Med, 88: 253–267, 2015.

[25] S. Petri, S. Körner, M. Kiaei, *"Nrf2/ARE signaling pathway: keymediator in oxidative stress and potential therapeutic target in ALS"*, Neurol Res Int, 2012: 878030, 2012.

[26] K. A. Jung and M. K. Kwak, *"The Nrf2 system as a potential target for the development of indirect antioxidants"*, Molecules, 15: 7266–7291, 2010.

[27] M. Kobayashi and M. Yamamoto, *"Molecular mechanisms activating the Nrf2-Keap1 pathway of antioxidant gene regulation"*, Antioxid Redox Signal, 7: 385–394, 2005.

[28] N. F. Villeneuve, A. Lau, D. D. Zhang, *"Regulation of the Nrf2-keap1 antioxidant response by the ubiquitin proteasome system: an insight into cullin-ring ubiquitin ligases"*, Antioxid Redox Signal, 13: 1699–1712, 2010.

[29] P. Milani, S. Gagliardi, E. Cova, C. Cereda, *"SOD1 transcriptional and posttranscriptional regulation and its potential implications in ALS"*, Neurol Res Int, 2011: 458427, 2011.

[30] P. Yenki, F. Khodagholi, F. Shaerzadeh, *"Inhibition of phosphorylation of JNK suppresses Abeta-induced ER stress and up regulates pro survival mitochondrial proteins in rat hippocampus"*, J Mol Neurosci, 49: 262–269, 2013.

[31] C. P. Ramsey, C. A. Glass, M. B. Montgomery et al., *"Expression of Nrf2 in neurodegenerative diseases"*, J Neuropathol Exp Neurol, 66: 75–85, 2007.

[32] G. Joshi and J. A. Johnson, *"The Nrf2-ARE pathway: a valuable therapeutic target for the treatment of neurodegenerative diseases"*, Recent Pat CNS Drug Discov, 7: 218–229, 2012.

[33] A. Wagenfeld, J. Gromoll, T. G. Cooper, *"Molecular cloning and expression of rat contraception associated protein 1 (CAP1), a protein putatively involved in fertilization"*, Biochem Biophys Res Commun 51: 545–549, 1998.

[34] V. Bonifati, P. Rizzu, M. J. van Baren et al., *"Mutations in the DJ-1 gene associated with autosomal recessive early-onset parkinsonism"*, Science, 299: 256–259, 2003.

[35] K. Honbou, N. N. Suzuki, M. Horiuchi et al., *"The crystal structure of DJ-1, a protein related to male fertility and Parkinson's disease"*, J Biol Chem, 278: 31380–31384, 2003.

[36] Q. Huai, Y. Sun, H. Wang et al., *"Crystal structure of DJ-1/RS and implication on familial Parkinson's disease"*, FEBS Lett, 549: 171–175, 2003.

[37] X. Tao and L. Tong, *"Crystal structure of human DJ-1, a protein associated with early onset Parkinson's disease"*, J Biol Chem, 278: 31372–31379, 2003.

[38] M. A. Wilson, J. L. Collins, Y. Hod, D. Ringe, G. A. Petsko, *"The 1.1- Å resolution crystal structure of DJ-1, the protein mutated in autosomal recessive early onset Parkinson's disease"*, Proc Natl Acad Sci U S A, 100: 9256–9261, 2003.

[39] M. Lakshminarasimhan, M. T. Maldonado, W. Zhou, A. L. Fink, M. A. Wilson, *"Structural impact of three Parkinsonism associated missense mutations on human DJ-1"*, Biochemistry, 47: 1381–1392, 2008.

[40] I. Irrcher, H. Aleyasin, E. L. Seifert et al., *"Loss of the Parkinson's disease-linked gene DJ-1 perturbs mitochondrial dynamics"*, Hum Mol Genet, 19: 3734–3746, 2010.

[41] N. J. Larsen, G. Ambrosi, S. J. Mullett, S. B. Berman, D. A. Hinkle, *"DJ-1 knock-down impairs astrocyte mitochondrial function"*, Neuroscience, 196: 251–264, 2011.

[42] S. J. Kim, Y. J. Park, I. Y. Hwang, M. B. Youdim, K. S. Park, Y. J. Oh, *"Nuclear translocation of DJ-1 during oxidative stress induced neuronal cell death"*, Free Radic Biol Med, 53: 936–950, 2012.

[43] L. Gu, T. Cui, C. Fan et al., *"Involvement of ERK1/2 signaling pathway in DJ-1-induced neuroprotection against oxidative stress"*, Biochem Biophys Res Commun, 383: 469–474, 2009.

[44] E. Andres-Mateos, C. Perier, L. Zhang et al., *"DJ-1 gene deletion reveals that DJ-1 is an atypical peroxiredoxin-like peroxidase"*, Proc Natl Acad Sci U S A, 104: 14807–14812, 2007.

[45] L. Gan, M. R. Vargas, D. A. Johnson, J. A. Johnson, *"Astrocyte-specific overexpression of Nrf2 delays motor pathology and synuclein aggregation throughout the CNS in the alpha-synuclein mutant (A53T) mouse model"*, J Neurosci, 32: 17775–17787, 2012.

[46] R. M. Canet-Avilés, M. A. Wilson, D. W. Miller et al., *"The Parkinson's disease DJ-1 is neuroprotective due to cysteine-sulfinic acid-driven mitochondrial localization"*, Proc Natl Acad Sci U S A, 101: 9103–9108, 2004.

[47] J. Waak, S. S. Weber, K. Gorner et al., *"Oxidizable residues mediating protein stability and cytoprotective interaction of DJ-1 with apoptosis signal-regulating kinase 1"*, J Biol Chem, 284: 14245–14257, 2009.

[48] M. A. Wilson, *"The role of cysteine oxidation in DJ-1 function and dysfunction"*, Antioxid Redox Signal, 15: 111–122, 2011.

[49] N. Zhong and J. Xu, *"Synergistic activation of the human MnSOD promoter by DJ-1 and PGC-1α: regulation by SUMOylation and oxidation"*, Hum Mol Genet, 17: 3357–3367, 2008.

[50] J. Y. Im, K. W. Lee, J. M. Woo, E. Junn, M. M. Mouradian, *"DJ-1 induces thioredoxin 1 expression through the Nrf2 pathway"*, Hum Mol Genet, 21: 3013–3024, 2012.

[51] H. Ren, K. Fu, D. Wang, C. Mu, G. Wang, *"Oxidized DJ-1 interacts with the mitochondrial protein BCL-XL"*, J Biol Chem, 286: 35308–35317, 2011.

[52] W. Zhou, M. Zhu, M. A. Wilson, G. A. Petsko, A. L. Fink, *"The oxidation state of DJ-1 regulates its chaperone activity toward α-synuclein"*, J Mol Biol, 356: 1036–1048, 2006.

[53] M. K. McCoy and M. R. Cookson, *"DJ-1 regulation of mitochondrial function and autophagy through oxidative stress"*, Autophagy, 7: 531–532, 2011.

[54] L. M. de Lau and M. M. Breteler, *"Epidemiology of Parkinson's disease"*, Lancet Neurol, 5: 525–535, 2006.

[55] G. Xiromerisiou, E. Dardiotis, V. Tsimourtou et al., *"Genetic basis of Parkinson disease"*, Neurosurg Focus, 28:E7.1–E7.7, 2010.

[56] H. Kumar, H.W. Lim, S. V. More et al., *"The role of free radicals in the aging brain and Parkinson's disease: convergence and parallelism"*, Int J Mol Sci, 13: 10478–10504, 2012.

[57] J. T. Greenamyre and T. G. Hastings, *"Parkinsons-divergent causes convergent mechanisms"*, Science, 304: 1120–1122, 2004.

[58] R. K. Chaturvedi and M. F. Beal, *"Mitochondrial diseases of the brain"*, Free Radic Biol Med, 63: 1–29, 2013.

[59] A. H. V. Schapira, A. Hartley, M. W. J. Cleeter, J. M. Cooper, *"Free radicals and mitochondrial dysfunction in Parkinson's disease"*, Biochem Soc Trans, 21: 367–370, 1993.

[60] A. Navarro, A. Boveris, M. J. Bàndez et al., *"Human brain cortex: mitochondrial oxidative damage and adaptive response in Parkinson disease and in dementia with Lewy bodies"*, Free Radic Biol Med, 46: 1574–1580, 2009.

[61] C. Buhmann, S. Arlt, A. Kontush et al., *"Plasma and CSF markers of oxidative stress are increased in Parkinson's disease and influenced by antiparkinsonian medication"*, Neurobiol Dis, 15: 160–170, 2004.

[62] C. Isobe, T. Abe, Y. Terayama, *"Levels of reduced and oxidized coenzyme Q-10 and 8-hydroxy-2'-deoxyguanosine in the cerebrospinal fluid of patients with living Parkinson's disease demonstrate that mitochondrial oxidative damage and/or oxidative DNA damage contributes to the neurodegenerative process"*, Neurosci Lett, 469: 159–163, 2010.

[63] S. Nikam, P. Nikam, S. K. Ahaley, A. V. Sontakke, *"Oxidative stress in Parkinson's disease"*, Indian J Clin Biochem, 24: 98–101, 2009.

[64] A. Jazwa, A. I. Rojo, N. G. Innamorato, M. Hesse, J. Fernàndez-Ruiz, A. Cuadrado, *"Pharmacological targeting of the transcription factor NRf2 at the basal ganglia provides disease modifying therapy for experimental parkinsonism"*, Antioxid Redox Signal, 14: 2347–2360, 2011.

[65] T. P. Williamson, D. A. Johnson, J. A. Johnson, *"Activation of the Nrf2-ARE pathway by siRNA knockdown of Keap1 reduces oxidative stress and provides partial protection from MPTP-mediated neurotoxicity"*, Neurotoxicology, 33: 272–279, 2012.

[66] Q. He, N. Song, F. Jia, H. Xu, X. Yu, J. Xie, H. Jiang, *"Role of alpha-synuclein aggregation and the nuclear factor E2-related factor 2/heme oxygenase-1 pathway in iron-induced neurotoxicity"*, Int J Biochem Cell Biol, 45: 1019–1030, 2013.

[67] M. C. Barone, G. P. Sykiotis, D. Bohmann, *"Genetic activation of Nrf2 signaling is sufficient to ameliorate neurodegenerative phenotypes in a Drosophila model of Parkinson's disease"*, Dis Model Mech, 4: 701–707, 2011.

[68] I. Lastres-Becker, A. Ulusoy, N. G. Innamorato, G. Sahin, A. Rabano, D. Kirik, A. Cuadrado, *"Alpha-synuclein expression and Nrf2 deficiency cooperate to aggravate protein*

aggregation, neuronal death and inflammation in early-stage Parkinson's disease", Hum Mol Genet, 21: 3173–3192, 2012.

[69] N. Mizushima and M. Komatsu, *"Autophagy: renovation of cells and tissues"*, Cell, 147: 728–741, 2011.

[70] B. Ravikumar, S. Sarkar, J. E. Davies et al., *"Regulation of mammalian autophagy in physiology and pathophysiology"*, Physiol Rev, 90: 1383–1435, 2010.

[71] R. A. Nixon, *"The role of autophagy in neurodegenerative disease"*, Nat Med, 19: 983–997, 2013.

[72] K. A. Malkus and H. Ischiropoulos, *"Regional deficiencies in chaperone-mediated autophagy underlie alpha-synuclein aggregation and neurodegeneration"*, Neurobiol Dis, 46: 732–744, 2012.

[73] X. Lin, L. Parisiadou, C. Sgobio, G. Liu et al., *"Conditional expression of Parkinson's disease-related mutant alpha-synuclein in the midbrain dopaminergic neurons causes progressive neurodegeneration and degradation of transcription factor nuclear receptor related 1"*, J Neurosci, 32: 9248–9264, 2012.

[74] S. J. Chinta, J. K. Mallajosyula, A. Rane, J. K. Andersen, *"Mitochondrial alpha-synuclein accumulation impairs complex I function in dopaminergic neurons and results in increased mitophagy in vivo"*, Neurosci Lett, 486: 235–239, 2010.

[75] M. Y. Aksenov, M. V. Aksenova, D. A. Butterfield, J. W. Geddes, W. R. Markesbery, *"Protein oxidation in the brain in Alzheimer's disease"*, Neuroscience, 103: 373–383, 2001.

[76] G. Benzi and A. Moretti, *"Age- and peroxidative stress-related modifications of the cerebral enzymatic activities linked to mitochondria and the glutathione system"*, Free Radic Biol Med, 19: 77–101, 1995.

[77] J. Y. Chan and M. Kwong, *"Impaired expression of glutathione synthetic enzyme genes in mice with targeted deletion of the Nrf2 basic-leucine zipper protein"*, Biochim Biophys Acta, 1517: 19–26, 2000.

[78] C. J. Harvey, R. K. Thimmulappa, A. Singh et al., *"Nrf2-regulated glutathione recycling independent of biosynthesis is critical for cell survival during oxidative stress"*, Free Radic Biol Med, 46: 443–453, 2009.

[79] Y. Namba, M. Tomonaga, H. Kawasaki et al., *"Apolipoprotein E immunoreactivity in cerebral amyloid deposits and neurofibrillary tangles in Alzheimer's disease and kuru plaque amyloid in Creutzfeldt-Jakob disease"*, Brain Res, 541: 163–166, 1991.

[80] M. K. Lee, D. R. Borchelt, G. Kim et al., *"Hyperaccumulation of FAD linked presenilin 1 variants in vivo"*, Nat Med, 3: 756–760, 1997.

[81] D. R. Marshak, S. A. Pesce, L. C. Stanley, W. S. Griffin, *"Increased S100 beta neurotrophic activity in Alzheimer's disease temporal lobe"*, Neurobiol Aging, 13: 1–7, 1992.

[82] H. M. Schipper, D. A. Bennett, A. Liberman et al.,*"Glial heme oxygenase-1 expression in Alzheimer disease and mild cognitive impairment"*. Neurobiol Aging, 27: 252–261, 2006.

[83] S. S. Shaftel, W. S. Griffin, M. K. O'Banion, *"The role of interleukin-1 in neuroinflammation and Alzheimer disease: an evolving perspective"*, J Neuroinflammation, 5: 7, 2008.

[84] M. A. Smith, P. L. Richey Harris, L. M. Sayre, J. S. Beckman, G. Perry, *"Widespread peroxynitrite mediated damage in Alzheimer's disease"*, J Neurosci, 17: 2653–2657, 1997.

[85] X. Wang, B. Su, G. Perry, M. A. Smith, X. Zhu, *"Insights into amyloid-beta-induced mitochondrial dysfunction in Alzheimer disease"*, Free Radic Biol Med, 43: 1569–1573, 2007.

[86] Y. Wang, K. Santa-Cruz, C. De Carli, J. A. Johnson, *"NAD(P)H:quinone oxidoreductase activity is increased in hippocampal pyramidal neurons of patients with Alzheimer's disease"*, Neurobiol Aging, 21: 525–531, 2000.

[87] A. K. Raina, D. J. Templeton, J. C. Deak, G. Perry, M. A. Smith, *"Quinone reductase (NQO1), a sensitive redox indicator, is increased in Alzheimer's disease"*, Redox Rep, 4: 23–27, 1999.

[88] D. A. Butterfield, J. Drake, C. Pocernich, A. Castegna, *"Evidence of oxidative damage in Alzheimer's disease brain: central role for amyloid beta-peptide"*, Trends Mol Med, 7: 548–554, 2001.

[89] W. R. Markesbery, *"Oxidative stress hypothesis in Alzheimer's disease"*, Free Radic Biol Med, 23: 134–147, 1997.

[90] M. A. Lovell, C. Xie, W. R. Markesbery, *"Acrolein is increased in Alzheimer's disease brain and is toxic to primary hippocampal cultures"*, Neurobiol Aging, 22: 187–194, 2001.

[91] W. R. Markesbery and M. A. Lovell, *"Four-hydroxynonenal, a product of lipid peroxidation, is increased in the brain in Alzheimer's disease"*, Neurobiol Aging, 19: 33–36, 1998.

[92] C. M. Lauderback, J. M. Hackett, F. F. Huang et al., *"The glial glutamate transporter, GLT-1, is oxidatively modified by 4-hydroxy-2-nonenal in the Alzheimer's disease brain: the role of Abeta1-42"*, J Neurochem, 78: 413–416, 2001.

[93] K. Kanninen, T. M. Malm, H. K. Jyrkkanen et al., *"Nuclear factor erythroid 2-related factor 2 protects against beta amyloid"*, Mol Cell Neurosci, 39: 302–313, 2008.

[94] R. Resende, P. I. Moreira, T. Proença, A. Deshpande et al., *"Brain oxidative stress in a triple-transgenic mouse model of Alzheimer disease"*, Free Radic Biol Med, 44: 2051–2057, 2008.

[95] C. Lee, G. H. Park, S. R. Lee, J. H. Jang, *"Attenuation of beta-amyloid-induced oxidative cell death by sulforaphane via activation of NF-E2-related factor 2"*, Oxid Med Cell Longev, 2013: 313510, 2013.

[96] X. H. Li, C. Y. Li, J. M. Lu, R. B. Tian, J. Wei, *"Allicin ameliorates cognitive deficits ageing-induced learning and memory deficits through enhancing of Nrf2 antioxidant signaling pathways"*, Neurosci Lett, 514: 46–50, 2012.

[97] C. J. Wruck, M. E. Gotz, T. Herdegen, D. Varoga, L. O. Brandenburg, T. Pufe, *"Kavalactones protect neural cells against amyloid beta peptide-induced neurotoxicity via extracellular signal-regulated kinase 1/2-dependent nuclear factor erythroid 2-related factor 2 activation"*, Mol Pharmacol, 73: 1785–1795, 2008.

[98] B. Eftekharzadeh, N. Maghsoudi, F. Khodagholi, *"Stabilization of transcription factor Nrf2 by tBHQ prevents oxidative stress-induced amyloid beta formation in NT2N neurons"*, Biochimie, 92: 245–253, 2010.

[99] H. V. Kim, H. Y. Kim, H. Y. Ehrlich, S. Y. Choi, D. J. Kim, Y. Kim, *"Amelioration of Alzheimer's disease by neuroprotective effect of sulforaphane in animal model"*, Amyloid, 20: 7–12, 2013.

[100] P. Mao, M. Manczak, M. J. Calkins et al., *"Mitochondria-targeted catalase reduces abnormal APP processing, amyloid beta production and BACE1 in a mouse model of Alzheimer's disease: implications for neuroprotection and lifespan extension"*, Hum Mol Genet, 21: 2973–2990, 2012.

[101] A. LoGerfo, L. Chico, L. Borgia, L. Petrozzi et al., *"Lack of association between nuclear factor erythroid-derived 2-like 2 promoter gene polymorphisms and oxidative stress biomarkers in amyotrophic lateral sclerosis patients"*, Oxid Med Cell Longev, 2014: 432626, 2014.

[102] P. Pasinelli and R. H. Brown, *"Molecular biology of amyotrophic lateral sclerosis: insights from genetics"*, Nat Rev Neurosci, 7: 710–723, 2006.

[103] M. Cozzolino, M. G. Pesaresi, V. Gerbino, J. Grosskreutz, M. T. Carri, *"Amyotrophic lateral sclerosis: new insights into underlying molecular mechanisms and opportunities for therapeutic intervention"*, Antioxid Redox Signal, 17: 1277–1330, 2012.

[104] S. Gagliardi, P. Milani, V. Sardone, O. Pansarasa, C. Cereda, *"From transcriptome to noncoding RNAs: implications in ALS mechanism"*, Neurol Res Int, 2012: 278725, 2012.

[105] L. Rossi, C. Valle, M. T. Carri, *"Altered gene expression, mitochondrial damage and oxidative stress: converging routes in motor neuron degeneration"*, Int J Cell Biol, 2012: 908724, 2012.

[106] M. J. Strong, *"The evidence for altered RNA metabolism in amyotrophic lateral sclerosis (ALS)"*, J Neurol Sci, 288: 1–12, 2010.

[107] R. J. Ferrante, S. E. Browne, L. A. Shinobu et al., *"Evidence of increased oxidative damage in both sporadic and familial amyotrophic lateral sclerosis"*, J Neurochem, 69: 2064–2074, 1997.

[108] S. C. Barber, R. J. Mead, P. J. Shaw, *"Oxidative stress in ALS: a mechanism of neurodegeneration and a therapeutic target"*, Biochim Biophys Acta, 1762: 1051–1067, 2006.

[109] S. C. Barber and P. J. Shaw, *"Oxidative stress in ALS: key role in motor neuron injury and therapeutic target"*, Free Radic Biol Med, 48: 629–641, 2010.

[110] K. Abe, L. H. Pan, M. Watanabe, T. Kato, Y. Itoyama, *"Induction of nitrotyrsone-like immunoreactivity in the lower motor neuron of amyotrophic lateral sclerosis"*, Neurosci Lett, 199: 152–154, 1995.

[111] M. F. Beal, R. J. Ferrante, S. E. Browne, R. T. Mathews et al., *"Increased 3-nitrotyrosine in both sporadic and familial amyotrophic lateral sclerosis"*, Ann Neurol, 42: 644–654, 1997.

[112] P. J. Shaw, P. G. Ince, G. Falkous, D. Mantle, *"Oxidative damage to protein in sporadic motor neuron disease spinal cord"*, Ann Neurol, 38: 691–695, 1995.

[113] R. G. Cutler, W. A. Pedersen, S. Camandola, J. D. Rothstein, M. P. Mattson, *"Evidence that accumulation of ceramides and cholesterol esters mediates oxidative stress—induced death of motor neurons in amyotrophic lateral sclerosis"*, Ann Neurol, 52: 448–457, 2002.

[114] W. A. Pedersen, W. Fu, J. N. Keller et al., *"Protein modification by the lipid peroxidation product 4-hydroxynonenal in the spinal cords of amyotrophic lateral sclerosis patients"*, Ann Neurol, 44: 819–824, 1998.

[115] R. G. Smith, Y. K. Henry, M. P. Mattson, S. H. Appel, *"Presence of 4-hydroxynonenal in cerebrospinal fluid of patients with sporadic amyotrophic lateral sclerosis"*, Ann Neurol, 44: 696–699, 1998.

[116] E. P. Simpson, Y. K. Henry, J. S. Henkel, R. G. Smith, S. H. Appel, *"Increased lipid peroxidation in sera of ALS patients: a potential biomarker of disease burden"*, Neurology, 62: 1758–1765, 2004.

[117] M. T. Carrì and M. Cozzolino, *"SOD1 and mitochondria in ALS: a dangerous liaison"*, J Bioenerg Biomembr, 43: 593–599, 2011.

[118] L. J. Martin, *"Mitochondrial pathobiology in ALS"*, J Bioenerg Biomembr, 43: 569–579, 2011.

[119] G. Manfredi and Z. Xu, *"Mitochondrial dysfunction and its role in motor neuron degeneration in ALS"*, Mitochondrion, 5: 77–87, 2005.

[120] D. Curti, A. Malaspina, G. Facchetti et al., *"Amyotrophic lateral sclerosis: oxidative energy metabolism and calcium homeostasis in peripheral blood lymphocytes"*, Neurology, 47: 1060–1064, 1996.

[121] J. Kong and Z. Xu, *"Massive mitochondrial degeneration in motor neurons triggers the onset of amyotrophic lateral sclerosis in mice expressing a mutant SOD1"*, J Neurosci, 18: 3241–3250, 1998.

[122] F. R. Wiedemann, G. Manfredi, C. Mawrin, M. F. Beal, E. A. Schon, *"Mitochondrial DNA and respiratory chain function in spinal cords of ALS patients"*, J Neurochem, 80: 616–625, 2002.

[123] E. Usarek, B. Gajewska, B. Kaźmierczak et al.,*"A study of glutathione S-transferase pi expression in central nervous system of subjects with amyotrophic lateral sclerosis using RNA extraction from formalin-fixed, paraffin-embedded material"*, Neurochem Res, 30: 1003–1007, 2005.

[124] M. Kuźma, Z. Jamrozik, A. Barańczyk-Kuźma, *"Activity and expression of glutathione S-transferase pi in patients with amyotrophic lateral sclerosis"*, Clin Chim Acta, 364: 217–221, 2006.

[125] S. Kato, M. Kato, Y. Abe et al., *"Redox system expression in the motor neurons in amyotrophic lateral sclerosis (ALS): immunohistochemical studies on sporadic ALS, superoxide dismutase 1 (SOD1)-mutated familial ALS, and SOD1-mutated ALS animal models"*, Acta Neuropathol, 110: 101–112, 2005.

[126] J. Kirby, E. Halligan, M. J. Baptista et al., *"Mutant SOD1 alters the motor neuronal transcriptome: implications for familial ALS"*, Brain, 128: 1686–1706, 2005.

[127] A. Sarlette, K. Krampfl, C. Grothe, N. V. Neuhoff, R. Dengler, S. Petri, *"Nuclear erythroid 2-related factor 2-antioxidative response element signaling pathway in motor cortex and spinal cord in amyotrophic lateral sclerosis"*, J Neuropathol Exp Neurol, 67: 1055–1062, 2008.

[128] T. Mimoto, K. Miyazaki, N. Morimoto et al., *"Impaired antioxydative Keap1/Nrf2 system and the downstream stress protein responses in the motor neuron of ALS model mice"*, Brain Res, 1446: 109–118, 2012.

[129] P. Milani, G. Ambrosi, O. Gammoh, F. Blandini, C. Cereda, *"SOD1 and DJ-1 converge at Nrf2 pathway: a clue for antioxidant therapeutic potential in neurodegeneration"*, Oxid Med Cell Longev, 2013: 836760, 2013.

[130] S. Kato, S. Horiuchi, J. Liu et al., *"Advanced glycation end product-modified superoxide dismutase-1 (SOD1)-positive inclusions are common to familial amyotrophic lateral sclerosis patients with SOD1 gene mutations and transgenic mice expressing human SOD1 with a G85R mutation"*, Acta Neuropathol, 100: 490–505, 2000.

[131] M. Neumann, D. M. Sampathu, L. K. Kwong et al., *"Ubiquitinated TDP-43 in frontotemporal lobar degeneration and amyotrophic lateral sclerosis"*, Science, 314: 130–133, 2006.

[132] I. R. Mackenzie, E. H. Bigio, P. G. Ince et al., *"Pathological TDP-43 distinguishes sporadic amyotrophic lateral sclerosis from amyotrophic lateral sclerosis with SOD1 mutations"*, Ann Neurol, 61: 427–434, 2007.

[133] T. J. Kwiatkowski Jr., D. A. Bosco, A. L. Leclerc et al.,*"Mutations in the FUS/TLS gene on chromosome 16 cause familial amyotrophic lateral sclerosis"*, Science, 323: 1205–1208, 2009.

[134] C. Vance, B. Rogelj, T. Hortobagyi et al.,*"Mutations in FUS, an RNA processing protein, cause familial amyotrophic lateral sclerosis type 6"*, Science, 323: 1208–1211, 2009.

[135] M. A. Gitcho, R. H. Baloh, S. Chakraverty et al., *"TDP-43 A315T mutation in familial motor neuron disease"*, Ann. Neurol, 63: 535–538, 2008.

[136] C. Lagier-Tourenne, M. Polymenidou, D. W. Cleveland, *"TDP-43 and FUS/TLS: emerging roles in RNA processing and neurodegeneration"*, Hum Mol Genet, 19: 46–64, 2010.

[137] T. Kanno, K. Tanaka, Y. Yanagisawa et al., *"A novel small molecule, N-(4-(2-pyridyl)(1,3-thiazol-2-yl))-2-(2,4,6-trimethylphenoxy) acetamide, selectively protects against oxidative stress-induced cell death by activating the Nrf2-ARE pathway: therapeutic implications for ALS"*, Free Radic Biol Med, 53: 2028–2042, 2012.

[138] R. J. Mead, A. Higginbottom, S. P. Allen et al., *"S[+] Apomorphine is a CNS penetrating activator of the Nrf2-ARE pathway with activity in mouse and patient fibroblast models of amyotrophic lateral sclerosis"*, Free Radic Biol Med, 61: 438–452, 2013.

[139] M. R. Vargas, D. A. Johnson, D. W. Sirkis et al., *"Nrf2 activation in astrocytes protects against neurodegeneration in mouse models of familial amyotrophic lateral sclerosis"*, J Neurosci, 28: 13574–13581, 2008.

[140] A. Nanou, A. Higginbottom, C. F. Valori et al., *"Viral delivery of antioxidant genes as a therapeutic strategy in experimental models of amyotrophic lateral sclerosis"*, Mol. Ther, 21: 1486–1496, 2013.

[141] M. R. Vargas, N. C. Burton, J. Kutzke et al., *"Absence of Nrf2 or its selective overexpression in neurons and muscle does not affect survival in ALS-linked mutant hSOD1 mouse models"*, PLoS One, 8: e56625, 2013.

[142] Y. Guo, Y. Zhang, D. Wen, W. Duan et al., *"The modest impact of transcription factor Nrf2 on the course of disease in an ALS animal model"*, Lab Invest, 93: 825–833, 2013.

[143] W. Duan, Y. Guo, H. Jiang, X. Yu, C. Li, *"MG132 enhances neurite outgrowth in neurons overexpressing mutant TAR DNA-binding protein-43 via increase of HO-1"*, Brain Res, 1397: 1–9, 2011.

[144] W. Duan, X. Li, J. Shi, Y. Guo, Z. Li, C. Li, *"Mutant TAR DNA-binding protein-43 induces oxidative injury in motor neuron-like cell"*, Neuroscience 169: 1621–1629, 2010.

[145] C. Stack, D. Ho, E. Wille, N. Y. Calingasan, C. Williams et al., *"Triterpenoids CDDO-ethyl amide and CDDO-trifluoroethyl amide improve the behavioral phenotype and brain pathology in a transgenic mouse model of Huntington's disease"*, Free Radic Biol Med, 49: 147–158, 2010.

[146] A. T. Dinkova-Kostova, W. D. Holtzclaw, R. N. Cole et al., *"Direct evidence that sulfhydryl groups of Keap1 are the sensors regulating induction of phase 2 enzymes that protect against carcinogens and oxidants"*, Proc Natl Acad Sci U S A, 99: 11908–11913, 2002.

[147] F. Hong, M. L. Freeman, D. C. Liebler, *"Identification of sensor cysteines in human Keap1 modified by the cancer chemopreventive agent sulforaphane"*, Chem Res Toxicol, 18: 1917–1926, 2005.

[148] J. H. Chen, H. P. Ou, C. Y. Lin, F. J. Lin et al., *"Carnosic acid prevents 6-hydroxydopamine-induced cell death in SH-SY5Y cells via mediation of glutathione synthesis"* Chem Res Toxicol, 25: 1893–1901, 2012.

[149] S. H. Koh, H. Kwon, K. S. Kim et al., *"Epigallocatechin gallate prevents oxidative-stress-induced death of mutant Cu/Znsuperoxide dismutase (G93A) motoneuron cells by alteration of cell survival and death signals"*, Toxicology, 202: 213–225, 2004.

[150] R. R. Hou, J. Z. Chen, H. Chen, X. G. Kang, M. G. Li, B. R. Wang, *"Neuroprotective effects of (-)-epigallocatechin-3- gallate (EGCG) on paraquat-induced apoptosis in PC12 cells"*, Cell Biol Int, 32: 22–30, 2008.

[151] S. H. Koh, S. M. Lee, H. Y. Kim et al., *"The effect of epigallocatechin gallate on suppressing disease progression of ALS model mice"*, Neurosci Lett, 395: 103–107, 2006.

[152] Z. Xu, S. Chen, X. Li, G. Luo, L. Li, W. Le, *"Neuroprotective effects of (-)-epigallocatechin-3-gallate in a transgenic mouse model of amyotrophic lateral sclerosis"*, Neurochem Res, 31: 1263–1269, 2006.

[153] H. K. Na and Y. J. Surh, *"Modulation of Nrf2-mediated antioxidant and detoxifying enzyme induction by the green tea polyphenol EGCG"*, Food Chem Toxicol, 46: 1271–1278, 2008.

[154] L. Romeo, M. Intrieri, V. D'Agata et al., *"The major green tea polyphenol, (-)-epigallocatechin-3-gallate, induces heme oxygenase in rat neurons and acts as an effective neuroprotective agent against oxidative stress"*, J Am Coll Nutr, 28: 492–499, 2009.

[155] V. Kesherwani, F. Atif, S. Yousuf, S. K. Agrawal, *"Resveratrol protects spinal cord dorsal column from hypoxic injury by activating Nrf-2"*, Neuroscience, 241: 80–88, 2013.

[156] A. M. Vincent, K. Kato, L. L. McLean, M. E. Soules, E. L. Feldman, *"Sensory neurons and Schwann cells respond to oxidative stress by increasing antioxidant defense mechanisms"*, Antioxid Redox Signal, 11: 425–438, 2009.

[157] H. Jiang, X. Tian, Y. Guo, W. Duan, H. Bu, C. Li, *"Activation of nuclear factor erythroid 2-related factor 2 cytoprotective signaling by curcumin protect primary spinal cord astrocytes against oxidative toxicity"*, Biol Pharm Bull, 34: 1194–1197, 2011.

[158] J. Wu, Q. Li, X. Wang et al., *"Neuroprotection by curcumin in ischemic brain injury involves the akt/nrf2 pathway"*, PLoS One, 8: e59843, 2013.

[159] T. G. Son, E. M. Kawamoto, Q. S. Yu, N. H. Greig, M. P. Mattson, S. Camandola, *"Naphthazarin protects against glutamate induced neuronal death via activation of the Nrf2/ARE pathway"*, Biochem Biophys Res Commun, 433: 602–606, 2013.

[160] R. Wang, J. Tu, Q. Zhang et al., *"Genistein attenuates ischemic oxidative damage and behavioral deficits via eNOS/Nrf2/HO-1 signaling"*, Hippocampus, 23: 634–647, 2013.

[161] T. Satoh, K. Kosaka, K. Itoh et al., *"Carnosic acid, a catechol type electrophilic compound, protects neurons both in vitro and in vivo through activation of the Keap1/Nrf2 pathway via S-alkylation of targeted cysteines on Keap1"*, J Neurochem, 104: 1116–1131, 2008.

[162] N. A. Kelsey, H. M. Wilkins, D. A. Linseman, *"Nutraceutical antioxidants as novel neuroprotective agents"*, Molecules, 15: 7792–7814, 2010.

[163] W. Ma, L. Yuan, H. Yu et al., *"Genistein as a neuroprotective antioxidant attenuates redox imbalance induced by β-amyloid peptides 25-35 in PC12 cells"*, Int J Dev Neurosci, 28: 289–295, 2010.

[164] Y. D. Xi, H. L. Yu, J. Ding et al., *"Flavonoids protect cerebrovascular endothelial cells through Nrf2 and PI3K from β- amyloid peptide-induced oxidative damage"*, Curr Neurovasc Res, 9: 32–41, 2012.

[165] N. A. Kaidery, R. Banerjee, L. Yang et al., *"Targeting Nrf2- mediated gene transcription by extremely potent synthetic triterpenoids attenuate dopaminergic neurotoxicity in the MPTP mouse model of Parkinson's disease"*, Antioxid Redox Signal, 18: 139–157, 2013.

[166] A. Neymotin, N. Y. Calingasan, E. Wille et al., *"Neuroprotective effect of Nrf2/ARE activators, CDDO ethylamide and CDDO trifluoroethylamide, in a mouse model of amyotrophic lateral sclerosis"*, Free Radic Biol Med, 51: 88–96, 2011.

[167] J. H. Lim, K. M. Kim, S. W. Kim, O. Hwang, H. J. Choi, *"Bromocriptine activates NQO1 via Nrf2-PI3K/Akt signaling: novel cytoprotective mechanism against oxidative damage"*, Pharmacol Res, 57: 325–331, 2008.

[168] S. Kalra, Y. Zhang, E. V. Knatko et al., *"Oral azathioprine leads to higher incorporation of 6-thioguanine in DNA of skin than liver: the protective role of the Keap1/Nrf2/ARE pathway"*, Cancer Prev Res, 4: 1665–1674, 2011.

Chapter 9

Nrf2 and Parkinson's Disease

Michael Todorovic and George D. Mellick

Additional information is available at the end of the chapter

http://dx.doi.org/10.5772/65450

Abstract

Parkinson's disease (PD) results from a complex interaction of environmental and genetic influences on a background of aging. Regardless of etiology, significant clinical advances rely on identifying the common biological pathways that underpin neuronal degeneration. Oxidative stress is consistently reported as a hallmark feature of PD. Recently, it has been demonstrated that Nrf2 modulation can protect neurons from parkinsonian agents and, in some instances, reverse motor symptoms of animal models. Furthermore, baseline aberrations of Nrf2 and its associated pathway have been reported in PD patients, and genetic variability—within and around the Nrf2 gene—may modify PD susceptibility and onset. Overall, Nrf2 dysregulation has been tentatively implicated in the pathogenesis of PD and may prove to be an effective therapeutic target.

Keywords: Parkinson's disease, Nrf2, oxidative stress, inflammation, dopamine

1. Introduction

Parkinson's disease (PD) is a neurodegenerative disorder characterized by a range of motor and nonmotor features. Clinically, PD diagnosis is based on the presence of distinctive cardinal motor features, including bradykinesia, resting tremor, postural instability, and rigidity [1]. Disease progression can be staged in accordance with developing neuropathological hallmarks that advance through presymptomatic and symptomatic phases [2]. The presymptomatic stages may last years to decades before the manifestation of classical PD-related motor symptoms [3–5]. Motor dysfunction is commonly associated with the loss of dopamine-producing neurons in the *substantia nigra pars compacta*, projecting throughout the nigrostriatal pathway. Further progressive and selective neuron loss will continue, ultimately culminating in a debilitating multisystem disorder [6].

Currently, the most efficacious medical treatments are limited to dopamine replacement therapies (levodopa), however, such medications wane in efficacy and can produce debilitating motor and nonmotor symptoms, prompting alternative approaches [7, 8]. Generally, PD medications agonize dopaminergic receptors, antagonize cholinergic receptors, and/or prolong dopamine activity (monoamine oxidase type-B inhibitors) [8]. While these approaches address symptoms, they provide no curative or disease-modifying effect, provoking researchers to isolate the pathogenic mechanism/s that underpin dopaminergic degeneration.

While the majority of PD cases are idiopathic (90–95%), etiological risk has been attributed to herbicide/pesticide exposure, heavy metals, rural living, aging, and genetic variability [9–11]. The role of these contributing factors in the pathogenesis of PD can be summarized as a complex interaction of environmental and genetic influences on a background of aging. Regardless of etiology, significant clinical advances rely on identifying the common biological pathways that underpin neuronal degeneration. Increasing evidence in this field suggests that oxidative stress is a major contributor in this process.

Free radicals, including reactive oxygen species (ROS) and reactive nitrogen species, are endogenous molecules, produced in cells as a by-product of metabolic systems (such as mitochondrial oxidative phosphorylation) and/or in response to an altered chemical environment. ROS are molecular species that contain an unpaired electron and are unstable and reactive. They are often implicated in disease (including numerous neurodegenerative diseases) and consist of hydroxyl, hydrogen peroxide, oxygen singlet, and superoxide radicals. Within a "steady-state" environment, the levels of ROS are often balanced by endogenous antioxidant defense mechanisms. However, if this balance is disrupted in the favor of ROS accumulation, a condition referred to as oxidative stress arises. Mitochondrial dysfunction, inflammation, and exercise are common endogenous generators of ROS. Environmental generators of ROS include cigarette smoke, pesticide exposure, and radiation. Overexposure to one or more of these factors may result in oxidative stress. Additionally, oxidative stress may occur if normal production of these reactive species cannot be appropriately managed. Thus, an inefficient antioxidant response mechanism may also result in increased risk for oxidative stress. An inability to balance redox systems and dispose of damaged cellular components may exacerbate ROS production and dramatically affect the survival of the cell through ROS-mediated lipid, protein, and DNA oxidation [12].

In general, neuronal cells are vulnerable to oxidative changes because of their high oxygen consumption and enrichment in fatty acids [13]. Furthermore, dopaminergic neurons are especially prone to oxidative-induced injury due to their capacity to produce ROS as a metabolic by-product. This can occur in two ways: (1) when dopamine is metabolized enzymatically (via monoamine oxidase) or (2) through auto-oxidation of dopamine (ultimately forming neuromelanin) [14]. Both means of processing dopamine produce ROS; enzymatic oxidation (1) forms hydrogen peroxide (H_2O_2), and nonenzymatic oxidation (2) produces superoxide (O^{2-}) and reactive quinones [15]. The production of ROS may be exacerbated by inflammation, neuronal damage, impaired mitochondrial management, and dysfunctional antioxidant response mechanisms [16]. However, the vast majority of ROS has been attributed to the mitochondria.

http://dx.doi.org/10.5772/65450

The major source of cellular energy (ATP) is produced by the mitochondria. This system is dependent on the simultaneous generation of a proton-motive force (termed the electron transport chain; ETC) across the mitochondrial inner membrane, driving the formation of ATP. The mitochondrial ETC came to the forefront of PD research after studies reported that 1-methyl-4-phenyl-1,2,3,6-tetrahydropyridine (MPTP), a mitochondrial ETC complex I inhibitor, induced Parkinsonian symptoms as a result of a toxic insult on nigrostriatal neurons [17]. Following these initial studies, complex I deficits have been identified not only in pharmacologically induced cases but also in idiopathic PD cases [18].

Endogenous antioxidants are required to maintain redox balance throughout the system. Many are transcriptionally regulated by the antioxidant "master regulator" Nrf2, which, upon direct stimulation from electrophilic compounds and ROS, will translocate to the nucleus and activate gene transcription of a gamut of antioxidant enzymes. Nrf2 supports and regulates, among others, the most abundant antioxidant family in the cell, glutathione. Interestingly, deficiency of reduced glutathione (GSH)—in the *substantia nigra* of PD brains—has consistently been reported [19, 20]. Recently, the Nrf2-mediated antioxidant response pathway has been directly and indirectly implicated in PD. This chapter will explore the role of Nrf2 and its pathway, in the development and management of PD, as stated within the contemporary literature.

2. The role of oxidative stress in Parkinson's disease

For years, markers of oxidative stress have been observed in postmortem brain tissue and disease models of PD. Currently it is unclear whether these factors are primary causes of PD or the result of established neurodegeneration. Regardless, studies have demonstrated that chronic ROS exposure can lead to the exacerbation of dopaminergic neuron death; implicating these processes in the pathophysiology of PD.

2.1. Mitochondria and oxidative stress

The brain is a major producer of ROS due to its extensive oxygen consumption; the nervous tissue is responsible for ~20% of total body oxygen consumption. The tightly regulated signaling systems of these neurons have a high energy demand provided primarily by the oxygen-dependent ATP production of the mitochondria. The ability of the mitochondria to fuse, divide, and migrate throughout the extended processes of neurons provide a dynamic adaptability in order to meet metabolic demands [21, 22]. Under normal resting conditions, mitochondria produce the ROS precursor molecule, superoxide ($O_2^{\bullet -}$), as a by-product of the electron transport chain. $O_2^{\bullet -}$ can produce hydrogen peroxide (H_2O_2) and hydroxyl ($^{\bullet}OH$); endogenous free radicals that can initiate lipid peroxidation. The brain is especially sensitive to lipid peroxidation due to the high concentration of polyunsaturated fatty acids. Because of this, neurons cells are extremely sensitive to the oxidative environment caused by mitochondrial defects; a number of these mitochondrial deficiencies have been implicated in neurodegenerative diseases, such as PD.

In the late 1970s, drug addicts presenting with Parkinsonian-like symptoms led to the discovery of MPTP, a contaminant of illicit meperidine synthesis (a synthetic analogue of heroin). Soon after its discovery, MPTP was used to produce a clinical phenotype, indistinguishable from PD, in primate species [23]. MPTP was shown to oxidize intraneuronally, the effect of which (underscored amongst other clinical hallmarks) was the selective destruction of dopaminergic neurons in the nigrostriatal system. More recently, a number of pesticides/herbicides have also demonstrated their efficacy as neurotoxins. Rotenone, a common herbicide, is often used in a manner similar to MPTP, to induce dopaminergic neurodegeneration in animal models. Both MPTP and rotenone, mechanistically, act upon the same pathway and inhibit complex I of the mitochondrial oxidative phosphorylation pathway [17]. Following these initial studies, complex I deficits have been identified not only in pharmacologically induced cases but also in idiopathic PD cases [18]. Furthermore, this reduction of complex I activity is not localized to the *substantia nigra,* but has been found in skeletal muscle, platelets, and fibroblasts of PD patients [24–27]. This evidence further implicates mitochondrial maintenance and subsequent free radical production in the pathogenesis of PD.

While the majority of PD is sporadic, rare genetic forms of the disease have been identified. So far, 23 chromosomal loci, termed the *PARK* loci, have been linked to both autosomal-dominant and autosomal-recessive inheritance patterns of PD. Interestingly, mutations within many of these loci have been associated with mitochondrial dysfunction and oxidative stress.

Dominantly inherited mutations in the alpha-synuclein (*SNCA*) gene were the first identified genetic forms of familial PD. Alpha-synuclein is the major proteinaceous constituent of the Lewy body, the key pathological hallmark of PD and other so-called synucleinopathies. This protein provided the first solid link between sporadic and familial forms of Parkinson's disease. Currently, studies suggest that, normally, alpha-synuclein plays a "protective role" as a chaperone, sequestering dysfunction proteins into aggregates [28–30]. However, these aggregates may lead to synaptic degeneration and ultimately cell death [31]. More recently, studies have suggested that alpha-synuclein aggregation can be induced by increased levels of ROS as a consequence of mitochondrial dysfunction, *in vitro* [32, 33]. Moreover, alpha-synuclein aggregation may further damage mitochondria, compounding the effects of oxidative stress [34].

Mutations within three *PARK* loci—*PARK2* (gene: *PARK2*; protein: Parkin), *PARK6* (gene: *PINK1*; protein: PINK1), and *PARK7* (gene: *PARK7*; protein: DJ-1)—are inherited in an autosomal recessive manner and have been linked to early-onset PD (i.e., symptoms present <45 years of age). Interestingly, animal genetic models have identified common and converging pathways for these gene products; these pathways focus on mitochondrial maintenance/dynamics, oxidative stress, and disrupted antioxidant pathways [35]. Individual disease models that knockout these genes result in increased susceptibility to H_2O_2 and excess ROS production [36–40]. Overall, converging evidence implicates certain *PARK* gene products in mitochondrial maintenance and ROS management. Subsequent dysfunction of these pathways therefore suggests that oxidative stress plays a central role in PD pathogenesis.

2.2. Dopamine metabolism and oxidative stress

As previously discussed, mitochondrial defects are associated with increased free radical production, and this has been theorized to play an important role in the pathogenesis of PD. However, these effects may be compounded by the ROS-enriched environment of highly metabolic dopamine-producing neurons. Dopamine can be metabolized enzymatically (via monoamine oxidase) or through auto-oxidation (ultimately forming neuromelanin) [14]. Both means of dopamine-processing produce ROS; enzymatic oxidation forms H_2O_2, and nonenzymatic oxidation produces O^{2-} and reactive quinones [15]. Production of ROS may be exacerbated by inflammation, neuronal damage, impaired mitochondrial management, and dysfunctional antioxidant response mechanisms [16].

Current evidence supports the hypothesis that PD is a consequence of genetic variation and environmental exposures, converging—ultimately—on oxidative stress. Therefore, it is important to characterize the role of antioxidant-response mechanisms, specifically Nrf2—commonly touted as the "master regulator" of oxidative stress—in the pathophysiology of PD.

3. Nrf2 in Parkinson's disease progression and pathology

Maintaining redox balance within an aging brain is reliant upon an efficient and an effective Nrf2-mediated pathway. However, aging appears to correlate with a decline in Nrf2 expression and transcriptional response, potentiating an individual's susceptibility to ROS accumulation [41, 42].

Nrf2 protein concentration, when isolated from the cerebral spinal fluid of PD patients with *LRRK2* mutations (G2019S), was positively, and significantly, associated with disease duration, motor scores, and the Unified Parkinson's Disease Rating Scale (UPDRS; well-established rating scale of Parkinson's disease symptom severity) [43]. This indicates that Nrf2 concentration (at least in the CSF) may increase with disease progression. Moreover, Nrf2 location may change in response to the oxidative profile of the cellular environment. This is also highlighted in a study that demonstrated that Nrf2 was found in the nucleus of nigral dopaminergic neurons in early Braak staging (1–2) PD patients, while a cytosolic localization was predominantly found for healthy, age-matched controls [44]. This translocation of Nrf2 to the nucleus in PD patients indicates an attempt to upregulate antioxidant responsive genes. Of these, the potent and diverse antioxidants, NQO1 and HO-1, are two that are strongly enhanced by Nrf2 activation. NQO1 is a successful metabolizer of dopamine-derived quinones [45] and has been reported at higher levels within the *subtantia nigra pars compacta* of PD patients compared to healthy controls [46]. NQO1 overexpression appears to protect cells against dopamine-mediated mitochondrial damage *in vitro* [47] and *in vivo* reduces MPTP toxicity [48]. Another Nrf2-transcribed antioxidant, HO-1, is also observed at higher concentrations in PD patients' blood serum compared to healthy controls [49]; the higher HO-1 concentrations were not observed in the blood serum of Alzheimer's disease patients, suggesting a disease-dependent Nrf2 recruitment. Unlike NQO1, no disease-specific differences for HO-1 expression have been observed in the *substantia nigra pars compacta* of PD patients. However, PD patients exhibiting

Lewy body pathology had a distinct HO-1 staining pattern within the periphery of these proteinaceous Lewy body inclusions [50]. This curious finding demonstrates the oxidative nature of Lewy bodies and the relationship between the Nrf2 pathway and PD pathology.

It is well established that the Nrf2 pathway is an integral player in the cellular response to the oxidative stress commonly associated with PD. It follows that a dysfunctional Nrf2 response may interfere with the normal healthy antioxidant management, and there is evidence that this contributes to risk for disease [51]. A number of studies have reported reduced antioxidant enzyme activity in the *substantia nigra pars compacta* of PD patient brains [52, 53]. Also, contemporary evidence suggests that genetic variability, in and around the Nrf2 encoding gene, is associated with disease susceptibility and modulates disease age-at-onset [11, 54, 55].

4. Nrf2 genetics and Parkinson's disease

Oxidative stress appears to lie at the nexus of genetic, pharmacologically induced, and idiopathic cases of Parkinsonism. Considering this, studies have begun to investigate the degree of influence that transcriptional "master regulators" of antioxidant response may impose on disease pathogenesis. A recent study has comprehensively screened *NFE2L2*, the gene that encodes Nrf2, for genetic sequence variants and correlated genetic variability with disease susceptibility [11]. Prior to this report, few candidate gene studies had investigated this relationship. A Taiwanese case–control study (PD = 480; controls = 526), which genotyped three Nrf2 promoter single nucleotide polymorphisms (SNPs), did not observe any significant individual polymorphism associations with PD [56]. Interestingly, a Polish case–control group reported that a specific haplotype (comprising these three promoter SNPs) was associated with disease protection and a delayed age-at-onset of PD. A further publication suggested that haplotypes of eight other SNPs found within and around *NFE2L2* altered disease risk and disease age-at-onset within two independent case–control groups (Polish and Swedish) [55]. This study was subsequently replicated in a European meta-analysis and in a larger Australian case–control study. The European case–control study (PD = 1038; controls = 1600) re-established the previously identified protective and disease-delaying Nrf2 promoter SNP haplotype and a number of individual polymorphisms associated with both earlier and delayed PD age-at-onset [54]. A large Australian study (PD = 1338; controls = 1379) further replicated the disease-delaying Nrf2 promoter haplotype and identified a SNP associated with an reduced disease risk [11]. Recently, two novel Nrf2-coding SNPs were identified and associated with PD within a Chinese population [57]. This study further demonstrated that overexpressing these alternate alleles reduced the expression of downstream Nrf2 products—glutathione s-transferase and HO-1. These studies provide compelling evidence that genetic variability within and around Nrf2 modulates PD risk and susceptibility. Due to the important role of Nrf2 as a functional respondent to oxidative threat, it is important to understand its influence on PD in the context of environmental exposures.

5. Nrf2 and environmental exposures in Parkinson's disease

A number of exogenous agents have shown to influence the development of PD; heavy metals (iron, copper, cadmium, manganese), insecticides/herbicides, and organic solvents are often reported in human epidemiological and animal studies [9, 10, 58]. Nrf2 stabilization from its constitutive repressor, KEAP1, and subsequent translocation to the nucleus are dependent upon exposure to electrophilic compounds and oxidative stress [59]. Some of the previously mentioned exogenous agents, implicated in PD, upregulate the Nrf2 signaling pathway as a response to mitigate potential damage [60]. The heavy metals, copper and iron, have been linked to oxidative stress and alpha-synuclein aggregation in PD. Experimental data have shown that accumulated ferrous iron downregulates Nrf2 and HO-1 expression, *in vitro*, promoting alpha-synuclein aggregation [61]. Furthermore, overexpression of HO-1 mitigates ferrous iron-induced cellular damage. Also, the ROS-mediated neurotoxic effects of excess copper exposure have shown to induce the Nrf2 pathway in zebrafish [62]. Acute cadmium exposure, tentatively associated with Parkinsonism [63], and manganese, known to produce Parkinson's-like motor dysfunction [64], both induce Nrf2 transcriptional activity *in vivo* [65].

Pesticide/herbicide exposures, classified as risk factors for PD, induce oxidative stress as a mechanism of neuronal cell death [66, 67]. The pesticide, deltamethrin, activates Nrf2 and downstream gene expression in rat brains [68]. Furthermore, Nrf2 activation protects neuronal cell lines from paraquat—a herbicide used to produce Parkinsonism in animal models. This data tentatively support the hypothesis that the Nrf2 pathway is modulated in response to PD-associated environmental exposures. Further studies have investigated whether genetic variability underlying Nrf2 and its various downstream products affect their cytoprotective activity in response to environmental insult.

One study observed that human olfactory neurosphere-derived cell lines carrying the minor allele of an Nrf2 SNP were significantly resilient to rotenone-induced cell death over a 5-day exposure [11]. In addition to Nrf2, PD risk from pesticide exposure has been associated, within certain populations, with genetic variability of Nrf2-transcribed genes. Individuals carrying *NQO1* SNPs are more susceptible to PD when exposed to pesticide compared to exposed individuals not carrying the variant [69]; while GST genotype has also shown to influence PD susceptibility upon exposure to paraquat [70, 71].

6. Nrf2 modulation as a neuroprotective strategy

As demonstrated in the literature, oxidative stress is highlighted as a major contributing factor in the pathogenesis of PD. Genetic, environmental, and idiopathic cases of PD have reported ROS imbalance, thereby tentatively implicating the "master regulator" of oxidative management, Nrf2, in the pathophysiological process. Numerous studies have decided to evaluate whether modulating Nrf2—either genetically or pharmacologically—influences disease susceptibility *in vitro* or *in vivo*. **Table 1** summarizes a number of these studies.

Target	Intervention	Model	Neuroprotective against:	References
Nrf2	Dimethyl fumarate (DMF)	Alpha-synuclein mouse model	Dopamine neuron loss in substantia nigra	[78]
	Overexpression	Mouse	H_2O_2 –and glutamate-treated cortical neurons	[79]
	Overexpression	Mixed primary neuron culture	Rotenone/ionomycin-induced cell death	[84]
	DMF[1,2]	Mouse[1] SH-SY5Y cells[2]	MPTP-induced neurotoxicity[1], 6-OHDA-treated cells[2]	[85][1], [86][2]
NQO1	Overexpression	SK-N-MC neuroblastoma	Dopamine-induced toxicity	[47]
Nrf[3,4], NQO1[9]	Triterpenoids[3] KMS04014[4,9]	C57B16 mouse[3,4,9]	MPTP-induced dopaminergic neuron toxicity[3,4,9]	[48][4], [87][3], [88][9]
Nrf2[5,6,10], NQO1[5,6], HO-1[5,6]	Bromocriptine[5], torularhodin[6], Selegiline[7]	PC12 cells[5,6,7]	H_2O_2-induced oxidative damage[5,6], MPP^+-induced oxidative damage[7]	[89][5], [90][6], [83][7]
Nrf2, NQO1, HO-1	α-Iso-cubebene	HT22 cells	Glutamate-induced oxidative damage	[91]
NQO1[8,9], HO-1[9]	KMS04014[8], isothiocyanate-3[9]	CATH.a cells[8,9], BV-2 cells[9]	H_2O_2/MPP^+–treated cells[8], lipopolysaccharide-treated cells[9]	[48][8], [88][9]

Superscript 1–9 denotes information obtained from a single source.

Table 1. Nrf2 and neuroprotection.

The ability of Nrf2 to attenuate disease relevant perturbations has been evaluated in neuronal cell line derivatives and various animal models. Many of these studies utilize a post-treatment strategy, perturbing the cells after Nrf2 is upregulated. Nrf2 activation can be influenced either pharmacologically (e.g., dimethyl fumarate (DMF), sulforaphane (SFN), or with tert-butyl hydroquinone (tBHQ)) or genetically. Perturbations are often performed with known Parkinsonian agents such as paraquat, rotenone, MPTP, 6-hydroxy dopamine (6-OHDA), or hydrogen peroxide (H_2O_2). This approach has consistently demonstrated that Nrf2 activation provides a successful neuroprotective strategy [72–77]. However, it must be noted that modulating Nrf2, prior to toxic treatment, does not reflect the insidious nature of PD and does not take into consideration the decades of accumulated cellular damage that has existed prior to clinical intervention. Notwithstanding this caveat, animal models of PD are providing compelling evidence that Nrf2 modulation offers significant protection against neuronal cell loss.

Current PD mouse models can recapitulate the histological hallmark of alpha-synuclein-containing aggregates and selective nigral dopaminergic neuron loss. The pharmacological targeting of Nrf2 in these models, via oral administration of DMF, can attenuate dopaminergic neuron loss in the *substantia nigra* [78]. Interestingly, this effect was not observed in Nrf2 knockout mice. Furthermore, Nrf2 knockout mice are more susceptible to cortical neuron cell damage caused by H_2O_2 and glutamate, while, on the other hand, they are significantly protected when Nrf2 is overexpressed [79]. Some studies have also shown that, in addition to Nrf2 overexpression, Keap1 repression may offer a successful strategy to restore neuron degeneration and motor dysfunction in an alpha-synuclein *Drosophila* model of PD [80].

To date, no studies have directly evaluated the efficacy of pharmacologically targeting Nrf2 as a treatment strategy for PD. Since the 1990s, the drug deprenyl (selegiline)—a type-B monoamine oxidase inhibitor (MAOI-B)—has been used as a pharmacological means to treat PD. The mechanism of action of MAOI-B is to inhibit the breakdown of monoamine neurotransmitters (such as dopamine). While this strategy temporally maintains synaptic dopamine concentrations, it also reduces the oxidative stress associated with dopamine metabolism [81]. While it has been known, since early in its use, that selegiline induced the expression of antioxidant enzymes [82], it has only recently been discovered that this is mediated by the activation of Nrf2 [83] and that activation of this pathway was sufficient to protect a neuronal-based cell line from oxidative damage.

7. Conclusion

Oxidative stress has been identifided as a major contributor in the pathogenesis of PD. Mechanisms of ROS production—contributing to the oxidative profile of neuronal cells—include mitochondrial respiration, dopamine metabolism, and environmental exposures. Normally, redox balance is managed by the transcription factor and antioxidant "master regulator," Nrf2. Studies have demonstrated that Nrf2 and its associated pathway products can be upregulated in PD patient brains; potentially, this highlights the body's attempt to mitigate oxidative stress. Furthermore, studies have also shown that reduced or dysfunctional Nrf2 can be found in PD tissue. Genetic variability within and around the Nrf2 gene has been associated with PD risk and age-at-onset, while genetic aberrations in Nrf2-mediated genes may influence an individual's risk of PD after exposure from environmental agents. Due to the functional role of Nrf2 in mitigating oxidative stress, many studies have investigated Nrf2 as a modulator of Parkinson's disease. Activation of Nrf2 attenuates neuronal damage caused by Parkinsonian agents, *in vitro* and *in vivo*. Overall, Nrf2 has been tentatively implicated in the pathophysiology of PD and may prove to be an effective therapeutic target.

Author details

Michael Todorovic and George D. Mellick*

*Address all correspondence to: g.mellick@griffith.edu.au

Eskitis Institute for Drug Discovery, Griffith University, Brisbane, Australia

References

[1] Jankovic, J., *Parkinson's disease: clinical features and diagnosis*. J Neurol Neurosurg Psychiatry, 2008. 79(4): p. 368–76.

[2] Braak, H., et al., *Staging of brain pathology related to sporadic Parkinson's disease*. Neurobiol Aging, 2003. 24(2): p. 197–211.

[3] Hawkes, C.H., K. Del Tredici, and H. Braak, *Parkinson's disease: the dual hit theory revisited*. Ann N Y Acad Sci, 2009. 1170: p. 615–622.

[4] Goldman, S.M., *A diagnostic algorithm for Parkinson's disease: what next?* Lancet Neurol, 2015. 14(10): p. 971–973.

[5] Haas, B.R., T.H. Stewart, and J. Zhang, *Premotor biomarkers for Parkinson's disease—a promising direction of research*. Transl Neurodegener, 2012. 1(1): p. 11.

[6] Braak, H., et al., *Pathology associated with sporadic Parkinson's disease—where does it end?* J Neural Transm Suppl, 2006. (70): p. 89–97.

[7] Siddiqi, S.H., et al., *The human experience with intravenous Levodopa*. Front Pharmacol, 2015. 6: p. 307.

[8] Connolly, B.S. and A.E. Lang, *Pharmacological treatment of Parkinson disease: a review*. JAMA, 2014. 311(16): p. 1670–1683.

[9] Montgomery, E.B., Jr., *Heavy metals and the etiology of Parkinson's disease and other movement disorders*. Toxicology, 1995. 97(1–3): p. 3–9.

[10] Tanner, C.M. and J.W. Langston, *Do environmental toxins cause Parkinson's disease? A critical review*. Neurology, 1990. 40(10 Suppl 3): p. suppl 17–30; discussion 30–1.

[11] Todorovic, M., et al., *Comprehensive assessment of genetic sequence variants in the antioxidant 'master regulator' nrf2 in idiopathic Parkinson's disease*. PLoS One, 2015. 10(5): p. e0128030.

[12] Young, I.S. and J.V. Woodside, *Antioxidants in health and disease*. J Clin Pathol, 2001. 54(3): p. 176–86.

[13] de Vries, H.E., et al., *Nrf2-induced antioxidant protection: a promising target to counteract ROS-mediated damage in neurodegenerative disease?* Free Radic Biol Med, 2008. 45(10): p. 1375–83.

[14] Sulzer, D., et al., *Neuromelanin biosynthesis is driven by excess cytosolic catecholamines not accumulated by synaptic vesicles*. Proc Natl Acad Sci U S A, 2000. 97(22): p. 11869–11874.

[15] Miyazaki, I. and M. Asanuma, *Dopaminergic neuron-specific oxidative stress caused by dopamine itself*. Acta Med Okayama, 2008. 62(3): p. 141–50.

[16] Lev, N., et al., *DJ-1 protects against dopamine toxicity.* J Neural Transm, 2009. 116(2): p. 151–160.

[17] Langston, J.W. and P. Ballard, *Parkinsonism induced by 1-methyl-4-phenyl-1,2,3,6-tetrahydropyridine (MPTP): implications for treatment and the pathogenesis of Parkinson's disease.* Can J Neurol Sci, 1984. 11(1 Suppl): p. 160–165.

[18] Schapira, A.H., et al., *Mitochondrial complex I deficiency in Parkinson's disease.* J Neurochem, 1990. 54(3): p. 823–827.

[19] Sian, J., et al., *Alterations in glutathione levels in Parkinson's disease and other neurodegenerative disorders affecting basal ganglia.* Ann Neurol, 1994. 36(3): p. 348–355.

[20] Pearce, R.K., et al., *Alterations in the distribution of glutathione in the substantia nigra in Parkinson's disease.* J Neural Transm, 1997. 104(6–7): p. 661–677.

[21] Bender, A., et al., *High levels of mitochondrial DNA deletions in substantia nigra neurons in aging and Parkinson disease.* Nat Genet, 2006. 38(5): p. 515–517.

[22] Court, F.A. and M.P. Coleman, *Mitochondria as a central sensor for axonal degenerative stimuli.* Trends Neurosci, 2012. 35(6): p. 364–372.

[23] Markey, S.P., et al., *Intraneuronal generation of a pyridinium metabolite may cause drug-induced parkinsonism.* Nature, 1984. 311(5985): p. 464–467.

[24] Parker, W.D., Jr., S.J. Boyson, and J.K. Parks, *Abnormalities of the electron transport chain in idiopathic Parkinson's disease.* Ann Neurol, 1989. 26(6): p. 719–723.

[25] Bindoff, L.A., et al., *Respiratory chain abnormalities in skeletal muscle from patients with Parkinson's disease.* J Neurol Sci., 1991. 104(2): p. 203–208.

[26] Krige, D., et al., *Platelet mitochondrial function in Parkinson's disease. The Royal Kings and Queens Parkinson Disease Research Group.* Ann Neurol, 1992. 32(6): p. 782–788.

[27] Wiedemann, F.R., et al., *Detection of respiratory chain defects in cultivated skin fibroblasts and skeletal muscle of patients with Parkinson's disease.* Ann N Y Acad Sci, 1999. 893: p. 426–429.

[28] Chandra, S., et al., *Double-knockout mice for alpha- and beta-synucleins: effect on synaptic functions.* Proc Natl Acad Sci U S A, 2004. 101(41): p. 14966–14971.

[29] Chandra, S., et al., *Alpha-synuclein cooperates with CSPalpha in preventing neurodegeneration.* Cell, 2005. 123(3): p. 383–396.

[30] Abeliovich, A., et al., *Mice lacking alpha-synuclein display functional deficits in the nigrostriatal dopamine system.* Neuron, 2000. 25(1): p. 239–252.

[31] Masliah, E., et al., *Dopaminergic loss and inclusion body formation in alpha-synuclein mice: implications for neurodegenerative disorders.* Science, 2000. 287(5456): p. 1265–1269.

[32] Esteves, A.R., et al., *Oxidative stress involvement in alpha-synuclein oligomerization in Parkinson's disease cybrids*. Antioxid Redox Signal, 2009. 11(3): p. 439–448.

[33] Pukass, K., O. Goldbaum, and C. Richter-Landsberg, *Mitochondrial impairment and oxidative stress compromise autophagosomal degradation of alpha-synuclein in oligodendroglial cells*. J Neurochem, 2015. 135(1): p. 194–205.

[34] Hsu, L.J., et al., *alpha-synuclein promotes mitochondrial deficit and oxidative stress*. Am J Pathol, 2000. 157(2): p. 401–410.

[35] Dodson, M.W. and M. Guo, *Pink1, Parkin, DJ-1 and mitochondrial dysfunction in Parkinson's disease*. Curr Opin Neurobiol, 2007. 17(3): p. 331–337.

[36] Gautier, C.A., T. Kitada, and J. Shen, *Loss of PINK1 causes mitochondrial functional defects and increased sensitivity to oxidative stress*. Proc Natl Acad Sci U S A, 2008. 105(32): p. 11364–11369.

[37] Clark, I.E., et al., *Drosophila pink1 is required for mitochondrial function and interacts genetically with parkin*. Nature, 2006. 441(7097): p. 1162–1166.

[38] Clements, C.M., et al., *DJ-1, a cancer- and Parkinson's disease-associated protein, stabilizes the antioxidant transcriptional master regulator Nrf2*. Proc Natl Acad Sci U S A, 2006. 103(41): p. 15091–15096.

[39] Krebiehl, G., et al., *Reduced basal autophagy and impaired mitochondrial dynamics due to loss of Parkinson's disease-associated protein DJ-1*. PLoS One, 2010. 5(2): p. e9367.

[40] Zhou, W., et al., *The oxidation state of DJ-1 regulates its chaperone activity toward alpha-synuclein*. J Mol Biol, 2006. 356(4): p. 1036–1048.

[41] Suh, J.H., et al., *Decline in transcriptional activity of Nrf2 causes age-related loss of glutathione synthesis, which is reversible with lipoic acid*. Proc Natl Acad Sci U S A, 2004. 101(10): p. 3381–3386.

[42] Shih, P.H. and G.C. Yen, *Differential expressions of antioxidant status in aging rats: the role of transcriptional factor Nrf2 and MAPK signaling pathway*. Biogerontology, 2007. 8(2): p. 71–80.

[43] Loeffler, D.A., et al., *CSF Nrf2 and HSPA8 in Parkinson's disease patients with and without LRRK2 gene mutations*. J Neural Transm (Vienna), 2016. 123(3): p. 179–187.

[44] Ramsey, C.P., et al., *Expression of Nrf2 in neurodegenerative diseases*. J Neuropathol Exp Neurol, 2007. 66(1): p. 75–85.

[45] Jia, Z., et al., *Dopamine as a potent inducer of cellular glutathione and NAD(P)H:quinone oxidoreductase 1 in PC12 neuronal cells: a potential adaptive mechanism for dopaminergic neuroprotection*. Neurochem Res, 2008. 33(11): p. 2197–2205.

[46] van Muiswinkel, F.L., et al., *Expression of NAD(P)H:quinone oxidoreductase in the normal and Parkinsonian substantia nigra*. Neurobiol Aging, 2004. 25(9): p. 1253–1262.

[47] Zafar, K.S., et al., *Overexpression of NQO1 protects human SK-N-MC neuroblastoma cells against dopamine-induced cell death*. Toxicol Lett, 2006. 166(3): p. 261–267.

[48] Son, H.J., et al., *Induction of NQO1 and neuroprotection by a novel compound KMS04014 in Parkinson's disease models*. J Mol Neurosci, 2015. 56(2): p. 263–272.

[49] Mateo, I., et al., *Serum heme oxygenase-1 levels are increased in Parkinson's disease but not in Alzheimer's disease*. Acta Neurol Scand, 2010. 121(2): p. 136–138.

[50] Schipper, H.M., A. Liberman, and E.G. Stopa, *Neural heme oxygenase-1 expression in idiopathic Parkinson's disease*. Exp Neurol, 1998. 150(1): p. 60–68.

[51] Cook, A.L., et al., *NRF2 activation restores disease related metabolic deficiencies in olfactory neurosphere-derived cells from patients with sporadic Parkinson's disease*. PLoS One, 2011. 6(7): p. e21907.

[52] Sofic, E., et al., *Antioxidant capacity in postmortem brain tissues of Parkinson's and Alzheimer's diseases*. J Neural Transm Suppl, 2006(71): p. 39–43.

[53] Perry, T.L. and V.W. Yong, *Idiopathic Parkinson's disease, progressive supranuclear palsy and glutathione metabolism in the substantia nigra of patients*. Neurosci Lett, 1986. 67(3): p. 269–74.

[54] von Otter, M., et al., *Genetic associations of Nrf2-encoding NFE2L2 variants with Parkinson's disease—a multicenter study*. BMC Med Genet, 2014. 15: p. 131.

[55] von Otter, M., et al., *Association of Nrf2-encoding NFE2L2 haplotypes with Parkinson's disease*. BMC Med Genet, 2010. 11: p. 36.

[56] Chen, Y.C., et al., *Genetic analysis of NFE2L2 promoter variation in Taiwanese Parkinson's disease*. Parkinsonism Relat Disord, 2013. 19(2): p. 247–250.

[57] Gui, Y., et al., *NFE2L2 variations reduce antioxidant response in patients with Parkinson disease*. Oncotarget, 2016. 7(10): p. 10756–10764.

[58] Menegon, A., et al., *Parkinson's disease, pesticides, and glutathione transferase polymorphisms*. Lancet, 1998. 352(9137): p. 1344–1346.

[59] Kobayashi, A., et al., *Oxidative and electrophilic stresses activate Nrf2 through inhibition of ubiquitination activity of Keap1*. Mol Cell Biol, 2006. 26(1): p. 221–229.

[60] Osburn, W.O. and T.W. Kensler, *Nrf2 signaling: an adaptive response pathway for protection against environmental toxic insults*. Mutat Res, 2008. 659(1–2): p. 31–39.

[61] He, Q., et al., *Role of alpha-synuclein aggregation and the nuclear factor E2-related factor 2/heme oxygenase-1 pathway in iron-induced neurotoxicity*. Int J Biochem Cell Biol, 2013. 45(6): p. 1019–1030.

[62] Jiang, W.D., et al., *Copper exposure induces oxidative injury, disturbs the antioxidant system and changes the Nrf2/ARE (CuZnSOD) signaling in the fish brain: protective effects of myo-inositol*. Aquat Toxicol, 2014. 155: p. 301–313.

[63] Okuda, B., et al., *Parkinsonism after acute cadmium poisoning*. Clin Neurol Neurosurg, 1997. 99(4): p. 263–265.

[64] Aschner, M., *Manganese: brain transport and emerging research needs*. Environ Health Perspect, 2000. 108 Suppl 3: p. 429–432.

[65] Casalino, E., et al., *The Nrf2 transcription factor contributes to the induction of alpha-class GST isoenzymes in liver of acute cadmium or manganese intoxicated rats: comparison with the toxic effect on NAD(P)H:quinone reductase*. Toxicology, 2007. 237(1–3): p. 24–34.

[66] Dick, S., et al., *Occupational titles as risk factors for Parkinson's disease*. Occup Med (Lond), 2007. 57(1): p. 50–56.

[67] Baltazar, M.T., et al., *Pesticides exposure as etiological factors of Parkinson's disease and other neurodegenerative diseases—a mechanistic approach*. Toxicol Lett, 2014. 230(2): p. 85–103.

[68] Li, H.Y., S.Y. Wu, and N. Shi, *Transcription factor Nrf2 activation by deltamethrin in PC12 cells: involvement of ROS*. Toxicol Lett, 2007. 171(1–2): p. 87–98.

[69] Fong, C.S., et al., *Pesticide exposure on southwestern Taiwanese with MnSOD and NQO1 polymorphisms is associated with increased risk of Parkinson's disease*. Clin Chim Acta, 2007. 378(1–2): p. 136–141.

[70] Dick, F.D., et al., *Gene-environment interactions in parkinsonism and Parkinson's disease: the Geoparkinson study*. Occup Environ Med, 2007. 64(10): p. 673–680.

[71] Goldman, S.M., et al., *Genetic modification of the association of paraquat and Parkinson's disease*. Mov Disord, 2012. 27(13): p. 1652–1658.

[72] Siebert, A., et al., *Nrf2 activators provide neuroprotection against 6-hydroxydopamine toxicity in rat organotypic nigrostriatal cocultures*. J Neurosci Res, 2009. 87(7): p. 1659–1669.

[73] Duffy, S., A. So, and T.H. Murphy, *Activation of endogenous antioxidant defenses in neuronal cells prevents free radical-mediated damage*. J Neurochem, 1998. 71(1): p. 69–77.

[74] Li, J., J.M. Lee, and J.A. Johnson, *Microarray analysis reveals an antioxidant responsive element-driven gene set involved in conferring protection from an oxidative stress-induced apoptosis in IMR-32 cells*. J Biol Chem, 2002. 277(1): p. 388–394.

[75] Li, J., et al., *Stabilization of Nrf2 by tBHQ confers protection against oxidative stress-induced cell death in human neural stem cells*. Toxicol Sci, 2005. 83(2): p. 313–328.

[76] Jazwa, A., et al., *Pharmacological targeting of the transcription factor Nrf2 at the basal ganglia provides disease modifying therapy for experimental parkinsonism*. Antioxid Redox Signal, 2011. 14(12): p. 2347–2360.

[77] Jakel, R.J., et al., *Nrf2-mediated protection against 6-hydroxydopamine*. Brain Res, 2007. 1144: p. 192–201.

[78] Lastres-Becker, I., et al., *Repurposing the NRF2 activator dimethyl fumarate as therapy against synucleinopathy in Parkinson's disease*. Antioxid Redox Signal, 2016. 25(2): p. 61–77.

[79] Kraft, A.D., D.A. Johnson, and J.A. Johnson, *Nuclear factor E2-related factor 2-dependent antioxidant response element activation by tert-butylhydroquinone and sulforaphane occurring preferentially in astrocytes conditions neurons against oxidative insult*. J Neurosci, 2004. 24(5): p. 1101–1112.

[80] Barone, M.C., G.P. Sykiotis, and D. Bohmann, *Genetic activation of Nrf2 signaling is sufficient to ameliorate neurodegenerative phenotypes in a Drosophila model of Parkinson's disease*. Dis Model Mech, 2011. 4(5): p. 701–707.

[81] Birkmayer, W., et al., *Implications of combined treatment with 'Madopar' and L-deprenil in Parkinson's disease. A long-term study*. Lancet, 1977. 1(8009): p. 439–443.

[82] Carrillo, M.C., et al., *The optimal dosage of (-)deprenyl for increasing superoxide dismutase activities in several brain regions decreases with age in male Fischer 344 rats*. Life Sci, 1993. 52(24): p. 1925–1934.

[83] Xiao, H., et al., *Deprenyl prevents MPP(+)-induced oxidative damage in PC12 cells by the upregulation of Nrf2-mediated NQO1 expression through the activation of PI3K/Akt and Erk*. Toxicology, 2011. 290(2–3): p. 286–94.

[84] Lee, J.M., et al., *NF-E2-related factor-2 mediates neuroprotection against mitochondrial complex I inhibitors and increased concentrations of intracellular calcium in primary cortical neurons*. J Biol Chem, 2003. 278(39): p. 37948–37956.

[85] Ahuja, M., et al., *Distinct Nrf2 signaling mechanisms of fumaric acid esters and their role in neuroprotection against 1-methyl-4-phenyl-1,2,3,6-tetrahydropyridine-induced experimental Parkinson's-like disease*. J Neurosci, 2016. 36(23): p. 6332–6351.

[86] Jing, X., et al., *Dimethyl fumarate attenuates 6-OHDA-induced neurotoxicity in SH-SY5Y cells and in animal model of Parkinson's disease by enhancing Nrf2 activity*. Neuroscience, 2015. 286: p. 131–140.

[87] Kaidery, N.A., et al., *Targeting Nrf2-mediated gene transcription by extremely potent synthetic triterpenoids attenuate dopaminergic neurotoxicity in the MPTP mouse model of Parkinson's disease*. Antioxid Redox Signal, 2013. 18(2): p. 139–157.

[88] Lee, J.A., et al., *A novel compound ITC-3 activates the Nrf2 signaling and provides neuroprotection in Parkinson's disease models*. Neurotox Res, 2015. 28(4): p. 332–345.

[89] Lim, J.H., et al., *Bromocriptine activates NQO1 via Nrf2-PI3K/Akt signaling: Novel cytoprotective mechanism against oxidative damage*. Pharmacol Res, 2008. 57(5): p. 325–331.

[90] Wu, J.L., et al., *Neuroprotective effects of torularhodin against H_2O_2-induced oxidative injury and apoptosis in PC12 cells*. Pharmazie, 2015. 70(1): p. 17–23.

[91] Park, S.Y., et al., *Neuroprotective effects of alpha-iso-cubebene against glutamate-induced damage in the HT22 hippocampal neuronal cell line*. Int J Mol Med, 2015. 35(2): p. 525–532.

Chapter 10

Oxidative Stress and Disease

Rosângela F.F de Araújo,
Danyelly Bruneska G. Martins and
Maria Amélia C.S.M. Borba

Additional information is available at the end of the chapter

http://dx.doi.org/10.5772/65366

Abstract

Typically in aerobic metabolism, organic compounds such as nucleic acids, proteins and lipids can undergo structural damage by oxidative reactions. This damage caused by reactive oxygen/nitrogen species has been recognized as "oxidative stress". Despite the biological systems present efficient enzymatic and nonenzymatic antioxidant systems, oxidative stress indicates a pro-oxidant/antioxidant imbalance in favor of excessive generation of free radicals or decrease in the removal rate. Various diseases such as cancer, diabetes, cardiovascular diseases and neurodegenerative clearly exemplify the chronic oxidative stress. Therefore, it is important to consider that at low and moderate ROS levels, it can, for example, act as signaling molecules that support cell proliferation and differentiation and activate survival pathways in response to stress. Correlations between oxidative stress and disease should be carefully investigated in order to understand whether oxidative stress actually increases susceptibility to a particular disease or opposite.

Keywords: oxidative stress, free radicals, oxidative damage, antioxidants, diseases

1. Introduction

The generation of free radicals is a continuous physiological process, fulfilling relevant biological functions. The mechanisms of generation of free radicals occur mostly in the mitochondria, cell membranes and cytoplasm. Reactive oxygen species (ROS) and reactive nitrogen species (RNS) are formed as unavoidable by-products of metabolism. During the metabolic processes, these radicals act as mediators for the transfer of electrons in various biochemical reactions. Its production, in appropriate proportions, is possible to generate adenosine triphosphate (ATP) through the electron transport chain; fertilization of the ovum; activation of genes and participation of defense mechanisms during the infection process [1]. The continuous production of free radicals during the metabolic processes culminated in the development of antioxidant defense mechanisms (enzymes and substances such as glutathione, metallothionein, vitamin A, vitamin C and vitamin E). These are intended to limit the intracellular levels of these reactive species and control the occurrence of damage caused by them. However, excessive production can lead to oxidative damage. The structural modifications in the molecules of nucleic acids, proteins and lipids caused by increased concentration of reactive oxygen species (ROS) and/or reactive nitrogen species (RNS) lead to various metabolic changes that may contribute to the development of neurological diseases, cardiovascular diseases, cancer, among others [2].

2. Oxidative stress and molecular damage

The installation process of oxidative stress arises from an imbalance between oxidants and antioxidants in favor of excessive generation of free radicals or removal speed thereof. This process leads to the oxidation of biomolecules with consequent loss of its biological functions and/or homeostatic imbalances, whose manifestation is the potential oxidative damage to cells and tissues. Accumulation of ROS/RNS can result in a number of deleterious effects such as lipid peroxidation, protein oxidation and DNA damage [3].

2.1. Nucleic acids damage

DNA and RNA are chemically unstable and vulnerable to hydrolysis, nonenzymatic methylation and oxidation, due to its susceptibility to endogenous and exogenous damage. The endogenous genotoxic agents are mainly produced by cellular metabolism and composed of ROS and RNS, estrogen metabolites and aldehydes produced by lipid peroxidation [4, 5].

There are two major endogenous oxidants causing nucleic acids damage: hydroxyl radicals ($HO^•$) and peroxynitrite (ONO_2^-). One major source of ROS is the mitochondrial respiration because up to 5% of oxygen undergoes single electron transfer and generates superoxide anion radical (O_2^-). The superoxide dismutase (SOD) converts O_2^- to hydrogen peroxide that should be reduced by catalase (CAT) or glutathione peroxidase (GPx), however when transition metals are present, it is reduced to hydroxyl radicals ($HO^•$). These radicals have a high reactivity, so it must be generated close to DNA or RNA in order to oxidize them. The generation of peroxynitrite (ONO_2^-) occurs by the reaction of nitric oxide (NO) and superoxide, both produced simultaneously in macrophages. Although these specimens can directly oxidize the nucleic acids, there is a secondary synergic mechanism of RNS to break the oxidative balance: the RNS are able to inhibit the enzyme FAPY glycosylase, a DNA repair mechanism to oxidation [6].

Oxidative stress can lead to different lesions in DNA, including direct modification of nucleotide bases, training sites apurinic/apyrimidinic, single strand break and much less frequently, breaking double strands. Considering all the bases of the nucleotides, guanine is most susceptible to oxidative changes because it has lower reduction potential and hydroxyl radicals interact with the imidazole ring of this nitrogenous base at positions C4, C5 and C8 [7].

The most studied marker for DNA oxidation is 8-hydroxydeoxyguanosine, a product of guanosine oxidation by $HO^{\bullet}$ [6, 8]. This product is able to pair with adenine, generating a GC/TA mutation upon replication [6]. It is also known that oxidative stress regulates DNA methylation, playing a role in epigenetics regulation. Epigenetics constitutes several mechanisms of controlling gene expression without changing DNA sequence, but responding fast and precisely to environmental changes. One of the most characterized methods of epigenetic regulation is DNA methylation. The methylation of DNA CpG islands is mediated by DNA methyltransferases (DNMTs), but when the ROS or RNS interacts with cytosine, it is chemically modified from 5-methylcytosine to 5-hydroxymethylcytosine, which prevents DNMT binding and alters methylation patterns [9].

For RNA oxidation, the most relevant marker is the homologue 8-hydroxyguanosine. It has been made clear that RNA is more often oxidized than DNA, due to its cellular location closer to ROS and RNS occurrence. The major consequences of RNA oxidization are the breakage of the strand and ribosomal dysfunction, preventing correct protein production [8].

2.2. Protein damage

The effects of oxidation in proteins can be observed in impaired protein folding, side-chain oxidation and backbone fragmentation, resulting in loss of function and stop a variety of biochemical processes. Among the amino acids, the cysteines and methionines are more easily oxidizable, but this reaction is reversible through disulfite reductases activity. However, the cysteine can also suffer irreversible oxidation reactions leading to the formation of S-carboxymethylcysteine and S-(2-Succinyl)cysteine, which implies the formation of fumarate and dicarbonyl groups covalently bound to cysteine residues. When the amino acids lysine, proline, arginine and threonine are oxidized, occurs the production of carbonyl derivatives, which are used as markers for oxidative stress. In the oxidation of aromatic amino acids, such as tyrosine, different products are formed due to interaction with ROS – dityrosine or RNS – 3-nitrotyrosine [8].

These oxidized-modified proteins are usually recognized and degraded in the cells, but some of them can accumulate over time and lead to cellular dysfunction. A physiological example is the lipofuscin, a brown-yellow pigment that is a product of iron-catalyzed oxidation (polymerization) of proteins and lipids, as it is extremely resistant to proteolysis, it accumulates and it is used as an aging marker [10].

2.3. Lipid damage

In biological systems, lipid peroxidation occurs in two forms, one enzymatically, involving the participation of cyclooxygenase and lipoxygenase in the oxidation of fatty acids and other nonenzyme medium, involving transition metal, the reactive species oxygen, nitrogen and others [11]. Excess peroxidation results are very damaging to the cell, despite contribute to the inflammatory response, due to its importance in the cascade reaction from arachidonic acid to prostaglandin formation. The action of free radicals on lipids leads to the formation of lipid hydroperoxides and aldehydes, such as malondialdehyde, 4-hydroxynonenal and isoprostanes that contribute further to increased cellular toxicity and can be detected in biological samples to measure oxidative stress. Lipid peroxidation disrupts the normal structure and function of lipid bilayers surrounding both the cell itself and in the membranes of organelles. In particular, the lipid peroxidation can alter membrane permeability, transportation and fluidity [12]. The chronicity of the process in question has important implications for the etiologic process of many chronic diseases, including atherosclerosis, diabetes, obesity, neurodegenerative disorders and cancer [1].

3. Antioxidant defense system

The antioxidant defense system has the primary objective to maintain the oxidative process within physiological limits and subject to regulation by preventing oxidative damage from spreading, culminating in systemic irreparable damage. The enzymatic defense system includes enzymes such as superoxide dismutase (SOD), catalase (CAT) and glutathione peroxidase (GPx). These enzymes act through mechanisms of preventing and/or controlling the formation of free radicals and species nonradical, involved with the initiation of chain reactions that culminate in propagation and process amplification and, consequently, the occurrence of oxidative damage. CAT and GPx enzymes act with the same purpose, to prevent the hydrogen peroxide accumulation. Such integrated action is of great importance, since this reactive species through the reactions of Fenton and Haber-Weiss, with the participation of iron and copper metals, culminates in the generation of $OH^{\bullet}$ radical against which there is no enzyme system defense [13, 14].

The human organism is constantly exposed to a vast number of molecules that can lead to oxidative stress, such as drugs and alcohol. However, there is a conserved cellular component to oxidative stress response, which is constituted by over 100 genes responsible for detoxification and antioxidant protein production. The first line of the antioxidant defense to exogenous toxins includes the enzymes involved in phase I and II metabolism. The phase I metabolism is responsible for increased compound polarity through oxidation, reduction or hydrolysis reactions. The phase II metabolism, in the other hand, is responsible for facilitating the cellular export of those compounds; its reactions are mainly glucuronidation, acetylation and sulfation [15].

The enzymes that compose the cytochrome P450 are the most responsible for oxidation of drugs, chemicals and various endogenous substrates, such as eicosanoids, cholesterol, vitamin D3 and arachidonic acid [16]. The P450 is a superfamily of heme-thiolated enzymes with over 2000 members [17]. In humans, 57 functional genes and 58 pseudogenes are grouped according to the sequence similarity in 18 families and 44 subfamilies. The CYP-enzymes that belong to the families 1, 2 and 3 are responsible for metabolizing up to 90% of the drugs, this phase I drug oxidation system is frequently redundant, but many drugs are metabolized to a clinical concentration by one or few CYPs only [18].

In steroidogenic tissues (converts cholesterol into pregnenolone via the P450 side chain cleavage enzyme) there is a prevalence of CYP450 enzymes located in mitochondria and the electron transport system is very susceptible to oxidative stress. During the electron transport, a leakage of electron to the ultimate acceptor leads to their binding to oxygen, being considered a primary source of ROS, this may result in acceleration of ROS production in mitochondria. In this context, it is considered the effectiveness of electron transfer from NADPH to CYP enzymes for monooxygenation of substrates as a source of ROS because during the uncoupling reaction, without the presence of any substrates, the electron-transfer chain oxidizes NADPH and yields ROS. During CYP2E1 metabolism is frequently observed this kind of uncoupling reactions, thus this enzyme is strongly associated to ROS production and oxidative stress [16]. The enzyme CYP2E1 is associated with the metabolism of small molecules, and can be induced by ethanol, obesity, diabetes and polyunsaturated fatty acids; this induction is related to toxicity and oxidative stress. Another mechanism of CYP2E1 activation is the reduction of glutathione levels, upon acetaminophen administration, for example. Besides, this drug increases lipid peroxidation and protein carbonylation, enhancing the ROS production due to higher activity of CYP2E1 and being associated to hepatotoxicity mediated by MAP-kinase pathway [16, 19].

Glutathione S-transferase (GST) is a family of intracellular enzymes that prevent the action of endogenous and exogenous toxins on the cells. GSTs are multifunctional enzymes that participate in the phase II of the xenobiotic metabolism and catalyze the nucleophilic attack of the reduced form of glutathione (GSH) to potentially hazardous compounds. How are involved in the metabolism of many carcinogens, environmental pollutants and cancer-fighting drugs, it is therefore reasonable to assume that the lack of specific isoenzymes has a significant effect on the tolerance of an organism to carcinogens [20]. Human GSTs are categorized into cytosolic/nuclear, mitochondrial and microsomal. Based on their amino acid sequences and/or nucleotide substrate specificity and immunological properties, seven classes of cytosolic GSTs are described: Alpha, Mu, Pi, Sigma, Theta, Omega and Zeta. Microsomal GSTs are designated MAPEG (membrane-associated proteins in eicosanoid and glutathione metabolism) and the only mitochondrial GST confirmed in humans is GST-kappa, which is also present in peroxisomes. GSTs are normally found in biological medium as homo or heterodimers and these dimers have two active sites whose activities are independent. After combining with reduced glutathione (GSH), these enzymes have higher specificity for a second substrate (the electrophilic). GST enzymes participate in the metabolism of endogenous and exogenous compounds, for example, polycyclic aromatic hydrocarbons, phenylalanine and tyrosine amino acids, testosterone and progesterone. These enzymes target endogenous compounds, maybe derived from peroxidation of polyunsaturated fatty acids present in cell membranes and the activity of reactive oxygen species [21–23].

4. Oxidative stress and neurological disorders

Conclusive evidence suggests that oxidative stress is a major contributor to the pathophysiology of a variety of neurodegenerative diseases, including Alzheimer's, Parkinson's, Huntington's, tardive dyskinesia (TD), epilepsy and acute diseases of the central nervous system, such as spinal cord injuries and/or brain traumatic. The human brain is vulnerable to oxidative stress due to many facts such as (i) metabolism of catecholamines; (ii) decrease in antioxidants; (iii) presence of transition metals; (iv) occurrence of brain trauma/injury; and also (v) the brain is a organ that proportionally requires more oxygen and (vi) expresses low levels of antioxidant enzymes, which contribute to formation of ROS. As a consequence of redox unbalance in brain, one of the most affected structures is the lipid membrane [24].

Parkinson's disease (PD) is characterized by loss of dopaminergic neurons in the substantia nigra pars compacta of the brain, leading to rigidity or slowing movements and postural instability. Most of the cases of PD are idiopathic and some cases are genetic-related, but in general context, aging is a determinant factor. In both idiopathic and genetic cases of PD, the oxidative stress plays a critical role in pathogenesis, being a common underlying mechanism. There is an elevated level of oxidized lipids, proteins and DNA associated with decreased glutathione level in the brain of PD patients. This increased susceptibility to oxidative damage in the dopaminergic neurons is due to (i) the presence of ROS generating enzymes, such as tyrosine hydroxylase and monoamine oxidase and (ii) these neurons contain iron, a catalyser of Fenton reaction ($Fe(II) + H_2O_2 \text{->} Fe(III) + OH^{\bullet} + OH^-$) that leads to superoxide radicals and hydrogen peroxide production [25].

A fact of Alzheimer's disease is the dysregulation of iron and copper homeostasis and various evidence of oxidative stress, mainly RNA oxidation. Neurons usually do not store big amounts of iron, but with aging there is an accumulation of iron in the brain, especially in microglia, astrocytes and neurons from cortex and hippocampus. If iron levels increase much more than ferritin, an iron-storage protein, it becomes free to catalyze Fenton's reaction [26].

The tardive dyskinesia (TD) is an adverse effect of antipsychotic use, it affects up to 25% of schizophrenic patients. However, as the majority of patients do not develop TD, it is considered that genetics factors may define its occurrence but TD pathophysiology remains unclear. One of the strongest hypotheses suggests that it is caused by oxidative stress originated from neurotoxic free-radical production upon antipsychotic medication. This affirmation is supported by genetic polymorphisms evaluated in genes that encode a mitochondrial enzyme that prevents oxidative damage due to energetic metabolism (manganese superoxide dismutase) and a cytosolic flavoenzyme that prevents quinone reduction (NADPH quinone oxidoreductase), playing a role in antioxidant defense [27].

5. Oxidative stress and metabolic syndrome

Metabolic syndrome is a term that designates a cluster of health problems often associated to modern life style, including obesity, insulin resistance, dyslipidemia, impaired glucose tolerance and high blood pressure. The metabolic syndrome is diagnosed when at least three of the following alterations are present: visceral obesity (waist circumference >102 cm in men or >88 cm in women); raised arterial blood pressure (>130/85 mm Hg); dysglycemia (fasting plasma glucose >100 mg dL); raised triglyceride concentrations (>150 mg dL) and low high-density lipoprotein (HDL) cholesterol (<40 mg dL in men or < 50 mg dL in women) [28].

The oxidative stress is related to metabolic syndrome in several ways: (i) H_2O_2 promotes insulin signaling, being associated with increased insulin resistance; (ii) superoxide anion is generated by angiotensin stimulation of NADPH and angiotensin II/angiotensin II type I receptor (AT1R), which plays a critical role in blood pressure control; (iii) hyperglycaemia leads to overproduction of superoxide by mitochondrial electron transfer chain, activating oxidative stress; (iv) elevated low-density lipoprotein (LDL) and low high-density lipoprotein (HDL) are correlated with oxidative stress and the dyslipidemia treatment with rosuvastatin is known to reduce oxidative stress through raise of antioxidant enzymes [28].

Due to oxidative DNA damage there is a direct correlation between diabetes and cancer. Diabetic patients present high levels of ROS because of elevated glucose, fatty acids and insulin blood levels; combined to lower antioxidative capacity derived from reduced glutathione synthesis. To support those findings, it has been proved that polymorphisms in peroxisome proliferator-activated receptor-γ coactivator-1α (PPARGC1A) – a protein that regulates mitochondrial electron transport, leads to decontrolled redox activity [29].

6. Oxidative stress and atherosclerosis

Atherosclerosis is defined as an arterial disease characterized by fibrous and cholesterol rich plaques. Atherosclerosis progression causes blood flow obstruction, hemorrhage due to rupture and thrombosis leading to strokes or myocardial infarctions. Many risk factors are associated with atherosclerosis development, the most widely known are serum low-density lipoprotein (LDL) cholesterol, low serum high-density lipoprotein (HDL) cholesterol, diabetes, hypertension, smoking, aging and oxidative stress [30].

During LDL oxidation, a progressive process and very important for the beginning of the formation of atheromatous plaque, the cholesterol is target of oxidants, which generate a variety of oxysterols. On the other hand, lipid peroxidation products (MDA and 4-HNE) can react with histidine, cysteine or lysine residues of proteins, leading to formation of stable Michael adducts with a hemiacetal structure or to Schiff bases that undergo a rearrangement generating the Amadori products. These aldehydes can derivatize Lys residues of apoB, which decreases the number of positive charges and interferes on LDL binding to LDLR and scavenger receptors [31].

In endothelial cells, besides stimulating the antioxidant defense (mainly by glutathione), Nrf2 (nuclear factor (erythroid-derived 2)-like 2) suppresses inflammation-associated expression of adhesion molecules and cytokines, which are associated with the early stage of atherogenesis [29]. NAD(P)H oxidases (NOXs) are major sources of ROS in the vasculature, producing superoxide from molecular oxygen using NAD(P)H as the electron donor and endothelial NO synthase (eNOS) produce NO which represents a key element in the vasoprotective function of the endothelium. However, pathological conditions associated with oxidative stress may become eNOS inefficient and promote the rapid inactivation of NO by excess superoxide [32].

There is growing evidence that reversal of oxidative stress with antioxidants can reduce the degree of myocardial ischemic injury and heart dysfunction [33].

7. Oxidative stress and infection

The pathological effects of NO and O_2^- in virus infection are in clear contrast to their beneficial antimicrobial effects in bacterial and fungal infections. In virus infections, NO and $ONOO^-$, which are primitive host-defense molecules, cause nonspecific oxidative damage in virus-infected tissue, leading to various pathological events. Virus-induced oxidative stress has been reported during HIV, influenza virus, HBV, hepatitis C virus, encephalomyocarditis virus (EMCV), respiratory syncytial virus (RSV), dengue virus (DENV) and others [34].

Studies including rotavirus-infected patients showed that viral infection stimulates NO production, decreases superoxide dismutase and glutathione peroxidase activities and increases inducible nitric oxide synthase (iNOS) mRNA and iNOS expression in murine ileum [35].

Influenza virus is probably the best characterized pathogen modulating redox homeostasis. Influenza-induced ROS production has been associated with host immune and inflammatory responses, as well as modulation of viral replication. Oxygen radicals and their derivatives are recognized as principal mediators of influenza virus-induced lung injury [36].

Within the Flaviviridae family, hepatitis C virus infection promotes oxidative stress and manipulates antioxidant systems, leading to liver damage and chronic disease. Elevated levels of reactive oxygen species (ROS) are considered as a major factor contributing to HCV-associated pathogenesis. HCV core protein is considered as a major regulator affecting the release of ROS from mitochondria. In this context, mitochondria play a crucial role for the production of ROS in HCV-infected cells. Several pathways are affected upon HCV infection to result in an induction of autophagy that interferes with various steps of the viral life cycle to promote a permanent viral infection. The assembly and release of viral particles are closely linked to the VLDL synthesis and occur via the secretory pathway. Elevated glucose production, enhanced fatty acid uptake or upregulation of genes involved in lipid and cholesterol synthesis may contribute to oxidative stress-induced insulin resistance linked to HCV infection [36].

Induction of iNOS and production of NO, accumulation of ROS and RNS, as well as perturbation of the reduced glutathione (GSH) content are all signatures of Dengue virus (DENV) infection in different human cells and animal models. DENV infection resulted in an intracellular accumulation of NAD(P)H oxidase (NOX2)-derived ROS in monocyte-derived dendritic cells (Mo-DCs). Alteration in the redox status of DENV-infected patients has been associated with increased inflammatory responses, cell death and correlated with different parameters associated with dengue disease [37].

The HPV infection, although necessary, is not sufficient to cause cancer and several studies have been devoted to the search for concurrent carcinogenic factors. Among these cofactors, many evidence support the role of ROS. It is clear that viral infection induces ROS that in turn causes damage to all types of biological macromolecules. Two different types of cooperative mechanisms are presumed to occur between ROS and HPV: (i) the ROS genotoxic activity and the HPV-induced genomic instability concur independently to the generation of the molecular damage necessary for the emergence of neoplastic clones. This first mode is merely a particular form of cocarcinogenesis and (ii) ROS specifically interacts with one or more molecular stages of neoplastic initiation and/or progression induced by the HPV infection [38, 39].

Therefore, it seems reasonable to hypothesize that, while in most cases the cells react to HPV infection and can overcome the virus-induced ROS by activating apoptosis leading to termination of viral replication and lesion regression, in some of the infected cells a steady state balance between ROS generation and detoxification is established, partly due to viral-induced antioxidant response. Thus, infected cells can aberrantly proliferate, paving the way to neoplastic progression HPV, exploit host cell survival mechanisms, through modulation of redox homeostasis, increasing the activity of catalase, SOD among other, as an adaptive response to the high ROS conditions of preneoplastic lesions. Elevated GST and GSH provide the HPV hosting cell with improved oxidative damage detoxifying systems, but suppression of p53 and iNOS together with induction of vascular endothelial growth factor (VEGF) and resistance to ROS leads to the suppression of apoptosis and generates an oxidant fitting cell phenotype. Therefore, the tumor cell adapts their metabolism in order to support their growth and survival, creating a paradox of high ROS production in the presence of high antioxidant levels [38, 39].

8. Oxidative stress and cancer

Many signaling pathways that regulate the metabolism of ROS are also linked to tumorigenesis [40, 41]. However, ROS can also promote tumor formation by inducing DNA mutations and pro-oncogenic signaling pathways. The production of low level of ROS is required for homeostatic signaling events. It can be driven by NAD(P)H and NAD(P)H oxidase (NOX), leading to the increase of cell proliferation and survival through the posttranslational modification of kinases and phosphatases. At moderate levels, ROS induce the expression of stress-responsive genes such as *HIF1A*, which in turn trigger the expression of proteins providing prosurvival signals, such as the glucose transporter GLUT1 (also known as SLC2A1) and vascular endothelial growth factor (VEGF). At low and moderate levels ROS can act as signaling molecules that sustain cellular proliferation and differentiation and activate stress-responsive survival pathways, stimulating the phosphorylation of protein kinase C (PKC), p38 mitogen-activated protein kinase (p38 MAPK), extracellular signal-regulated kinase (ERK)1/2, phosphoinositide 3-kinase/serine-threonine kinase (PI3K/Akt), protein kinase B (PKB) and JUN N-terminal kinase (JNK) [40, 42].

The regulation of oxidative stress is an important factor not only for tumor development but also for the responses to anticancer therapies. As high ROS levels are harmful to cells, oxidative stress can have a tumor-suppressive effect. This imparts pressure on cancer cells to adapt by developing strong antioxidant mechanisms. But despite having an enhanced antioxidant system, cancer cells maintain higher ROS levels than normal cells. At high levels, ROS can cause damage to macromolecules, including DNA; induce the activation of protein kinase Cδ (PKCδ), triggering senescence; and/or cause permeabilization of the mitochondria, leading to the release of cytochrome *c* and apoptosis. ROS are also involved in the increased expression of antioxidant genes related to the activation of transcription factors such as the Nrf2, activator protein 1 (AP-1), nuclear factor kB (NF-kB) and p53 [40–42].

The role of ROS in carcinogenic process can be either pro or anti oncogenic, and it can be summarized as follows: (i) regulating tumor development and signaling pathways for cell progression through ERK1/2 activation and ligand-independent RTK activation; (ii) regulating chronic inflammation for example through NF-kB activation; (iii) controlling tumor suppressor expression and cell cycle inhibitors; (iv) mediating angiogenesis by the release of vascular endothelial growth factor (VEGF) and angiopoietin; (v) favoring metastasis and tissue invasion due to metalloproteinase secretion; (vi) avoiding cellular death by activating SRC and PI3K/AKT pathway. Additionally, generating ROS is the mechanism of attack used by most of chemotherapies and radiotherapy [43, 44].

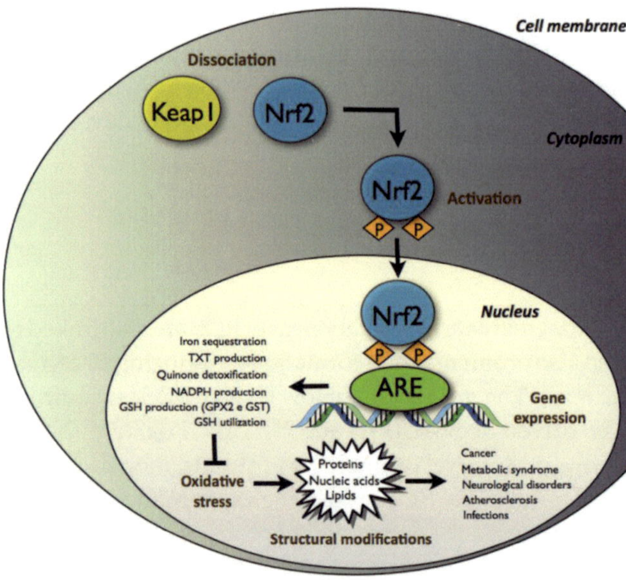

Figure 1. Keap1 (Kelch-like ECH-associated protein 1) sequesters Nrf2 (nuclear factor erythroid-derived 2) in the cytoplasm by binding to its aminoterminal regulatory domain. Keap1 is a sulfhydryl (S)-rich protein, and several cysteine residues mediate the Keap1–inducer interaction. When the interaction between Keap1 and Nrf2 disrupts, it allows Nrf2 to translocate to the nucleus. In the nucleus, Nrf2 controls several different antioxidant pathways by activating the expression of GSTs and other genes. This control is important to avoid cellular wear caused by oxidative stress, thus hindering the onset of various diseases.

The interindividual variation of the activity of antioxidant enzymes, for example, GST, considered by both environmental factors (e.g., diet and exposure to toxins such as cigarette) and genetic, is directly related to the etiology of cancer. Cytosolic GST present polymorphisms in humans and, this is probably the cause for differences in interindividual response to xenobiotics. The first studies in this area have addressed the correlation between GSTM1 null and/or GSTT1 null genotypes and a higher incidence of lung cancer, bladder, breast, colorectal head/neck. The discovery of allelic variants of GSTP1, encoding enzymes with reduced catalytic activity, led many researchers to examine the hypothesis that the combinations of polymorphisms of the Mu class, Pi and Theta of GST contribute to disorders with environmental factors [45, 46]. Studies with mice that exhibited a homozygous deletion of Nrf2 showed that Nrf2 is critical for inducing hepatic glutathione S-transferase (GST), NAD(P)H: quinone oxidoreductase (NQO1) and regulating levels of glutathione (**Figure 1**) [47].

Besides genetic variants of GST, changes in phase I enzyme activity as encoded by the cytochrome P450 family can also have implications for the metabolism of specific nitrosamines from the tobacco, alcohol and other carcinogenic substances [48].

The GST enzymes are part of an integrated protection system, so it is important to note that the efficiency of this system depends on the combined action of other enzymes, such as γ-glutamylcysteine synthase γGluCysS) and glutathione synthase, in order to provide glutathione as well as carriers to facilitate the elimination of glutathione conjugates [21].

9. Conclusion

The modulation of intracellular ROS levels is crucial for cellular homeostasis, and different ROS levels can induce different biological responses. It can occur due to the accumulation of intrinsic and/or environmental factors, such as hypoxia, enhanced cellular metabolic activity, mitochondrial dysfunction, increased activity of oxidases, lipoxygenases and cyclooxygenases. The accumulation of free radicals can lead to important changes in the structure of nucleic acids, proteins and lipids, altering their functions with consequent impact on cellular metabolism. These changes create conditions favorable to the onset of different diseases. The determination of oxidative stress markers and plasma antioxidants can suggest a targeted therapy against deficiencies in cell protection systems and it could be useful in an attempt to minimize complications caused by increased oxidative stress, leading to a better prognosis of various diseases.

Author details

Rosângela F.F de Araújo[1,2*], Danyelly Bruneska G. Martins[1,2] and
Maria Amélia C.S.M. Borba[2]

*Address all correspondence to: rfrade@prospecmol.org

1 Department of Biochemistry – Federal University of Pernambuco, Recife, Brazil

2 Immunopathology Keizo Asami Laboratory – Federal University of Pernambuco, Recife, Brazil

References

[1] Lobo V, Patil A, Phatak A, Chandra N. Free radicals, antioxidants and functional foods: impact on human health. Pharmacognosy Reviews. 2010;4(8):118–126. DOI: 10.4103/0973-7847.70902.

[2] Rahman K. Studies on free radicals, antioxidants, and co-factors. Clinical Interventions in Aging. 2007;2(2):219–236. PMCID: PMC2684512.

[3] Lushchak VI. Free radicals, reactive oxygen species, oxidative stress and its classification. Chemico-Biological Interactions 2014;224:164–175. DOI: 10.1016/j.cbi.2014.10.016

[4] Ranchoux B, Meloche J, Paulin R, Boucherat O, Provencher S, Bonnet S. DNA damage and pulmonary hypertension. International Journal of Molecular Sciences. 2016;17(6): 990. DOI: 10.3390/ijms17060990

[5] Marengo B, Nitti M, Furfaro AL, Colla R, De Ciucis C, Marinari UM, et al. Redox homeostasis and cellular antioxidant systems: crucial players in cancer growth and therapy. Oxidative Medicine and Cellular Longevity 2016;2016:1–16. DOI: 10.1155/2016/6235641

[6] De Bont R, van Larebeke N. Endogenous DNA damage in humans: a review of quantitative data. Mutagenesis. 2004;19(3):169–185. DOI: 10.1093/mutage/geh025

[7] Smith JA, Park S, Krause JS, Banik NL. Oxidative stress, DNA damage, and the telomeric complex as therapeutic targets in acute neurodegeneration. Neurochemistry International. 2013; 62: 764–775. DOI:10.1016/j.neuint.2013.02.013.

[8] Pisoschi AM, Pop A. The role of antioxidants in the chemistry of oxidative stress: a review. European Journal of Medicinal Chemistry 2015;97:55–74. DOI: 10.1016/j.ejmech. 2015.04.040

[9] Zhao H, Han Z, Ji X, Luo Y. Epigenetic regulation of oxidative stress in ischemic stroke. Aging and Disease. 2016;7(3):295–306. DOI: 10.14336/AD.2015.1009

[10] Amir Aslani B, Ghobadi S. Studies on oxidants and antioxidants with a brief glance at their relevance to the immune system. Life Sciences 2016;146:163–173. DOI: 10.1016/ j.lfs.2016.01.014

[11] Ayala A, Muñoz MF, Argüelles S. Lipid peroxidation: production, metabolism, and signaling mechanisms of malondialdehyde and 4-hydroxy-2-nonenal. Oxidative Medicine and Cellular Longevity. 2014; 2014:1–31. DOI: 10.1155/2014/360438

[12] Morita M, Naito Y, Yoshikawa T, Niki E. Plasma lipid oxidation induced by peroxynitrite, hypochlorite, lipoxygenase and peroxyl radicals and its inhibition by antioxidants as assessed by diphenyl-1-pyrenylphosphine. Redox Biology 2016;8:127–135. DOI: 10.1016/j.redox.2016.01.005

[13] Schneider CD, Oliveira AR. Free radicals of oxygen and exercise: mechanisms of formation and adaptation to physical training. Revista Brasileira de Medicina do Esporte. 2004;10(4). DOI: 10.1590/S1517-86922004000400008

[14] Uhl L, Gerstel A, Chabalier M, Dukan S. Hydrogen peroxide induced cell death: one or two modes of action? Heliyon. 2015;1(4). DOI: 10.1016/j.heliyon.2015.e00049

[15] Jacob C, Winyard P. Redox Signaling and Regulation in Biology and Medicine. Wiley-VCH. 2009. Weinheim, Germany. DOI: 10.1002/9783527627585

[16] Bhattacharyya S, Sinha K, Sil PC. Cytochrome P450s: mechanisms and biological implications in drug metabolism and its interaction with oxidative stress. Current Drug Metabolism 2014;15:719–742. DOI: 10.2174/1389200215666141125121659

[17] Lewis DF Human cytochromes P450 associated with the phase 1 metabolism of drugs and other xenobiotics: a compilation of substrates and inhibitors of the CYP1, CYP2 and CYP3 families. Current Medicinal Chemistry. 2003;10(19):1955–1972. DOI: 10.2174/0929867033456855

[18] Zanger UM, Schwab M. Cytochrome P450 enzymes in drug metabolism: regulation of gene expression, enzyme activities, and impact of genetic variation. Pharmacology & Therapeutics. 2013;138(1):103–141. DOI: 10.1016/j.pharmthera.2012.12.007

[19] Tanaviyutpakdee P, Yoovathaworn K, Sirivarasai J, Chanprasertyothin S. Role of CYP2E1 and NQO1 polymorphisms in oxidative stress derived cancer in Thais with and without dyslipidemia. Asian Biomedicine. 2015;9(5):601–611. DOI: 10.5372/1905-7415.0904.430

[20] Zheng W, Wen W-Q, Gustafson DR, Gross M, Cerhan JR, Folsom AR. GSTM1 and GSTT1 polymorphisms and postmenopausal breast cancer risk. Breast Cancer Research and Treatment. 2002 Jul;74(1):9–16. DOI: 10.1023/A:1016005100958

[21] Huber PC, Almeida WP, Fátima Â de. Gluthathione and related enzymes: the biological role and importance in pathological process. Quimica Nova. 2008;31(5):1170–1179. DOI: 10.1590/S0100-40422008000500046

[22] Lo H-W, Ali-Osman F. Genetic polymorphism and function of glutathione S-transferases in tumor drug resistance. Current Opinion in Pharmacology. 2007;7(4):367–374. DOI: 10.1016/j.coph.2007.06.009

[23] Sheehan D, Meade G, Foley VM, Dowd CA. Structure, function and evolution of glutathione transferases: implications for classification of non-mammalian members of an ancient enzyme superfamily. The Biochemical Journal. 2001 Nov 15;360(Pt 1):1–16. DOI: 10.1042/0264-6021:3600001

[24] Rao AV, Balachandran B. Role of oxidative stress and antioxidants in neurodegenerative diseases. Nutritional Neuroscience. 2002;5: 291–309. DOI: 10.1080/1028415021000033767

[25] Hwang O. Role of oxidative stress in Parkinson's Disease. Experimental Neurobiology. 2013;22(1):11–17. DOI: 10.3233/JPD-130230

[26] Hofer T, Perry G. Nucleic acid oxidative damage in Alzheimer's disease-explained by the hepcidin-ferroportin neuronal iron overload hypothesis? Journal of Trace Elements in Medicine and Biology: Organ of the Society for Minerals and Trace Elements (GMS). 2016; in press. DOI:10.1016/j.jtemb.2016.06.005

[27] Cho C-H, Lee H-J. Oxidative stress and tardive dyskinesia: Pharmacogenetic evidence. Progress in Neuro Psychopharmacology and Biological Psychiatry. 2013 Oct 1; 46:207–13. DOI: http://dx.doi.org/10.1016/j.pnpbp.2012.10.018

[28] Bonomini F, Rodella LF, Rezzani R. Metabolic syndrome, aging and involvement of oxidative stress. Aging and Disease. 2015;6(2):109. DOI: 10.14336/AD.2014.0305

[29] Lee SC, Chan JC. Evidence for DNA damage as a biological link between diabetes and cancer. Chinese Medical Journal. 2015;128(11):1543–1548. DOI: 10.4103/0366

[30] Mimura J, Itoh K. Role of Nrf2 in the pathogenesis of atherosclerosis. Free Radical Biology and Medicine 2015;88:221–232. DOI: 10.1016/j.freeradbiomed.2015.06.019

[31] Salvayre R, Negre-Salvayre A, Camaré C. Oxidative theory of atherosclerosis and antioxidants. Biochimie 2016;125:281–296. DOI: 10.1016/j.biochi.2015.12.014

[32] Li H, Horke S, Förstermann U, Zhang DX, Gutterman DD, Stocker R, et al. Vascular oxidative stress, nitric oxide and atherosclerosis. Atherosclerosis. 2014 Nov;237(1):208–219. DOI: 10.1016/j.atherosclerosis.2014.09.001

[33] Wang D, Wang J, Liu Y, Zhao Z, Liu Q. Roles of Chinese herbal medicines in ischemic heart diseases (IHD) by regulating oxidative stress. International Journal of Cardiology 2016 Oct;220:314–319. DOI: 10.1016/j.ijcard.2016.06.161

[34] Akaike T, Maeda H. Nitric oxide and virus infection. Immunology. 2000 Nov;101(3):300–308. DOI: 10.1046/j.1365-2567.2000.00142.x

[35] Guerrero CA, Acosta O. Inflammatory and oxidative stress in rotavirus infection. World Journal of Virology. 2016 May 12;5(2):38–62. DOI: 10.5501/wjv.v5.i2.38

[36] Medvedev R, Ploen D, Hildt E, Medvedev R, Ploen D, Hildt E. HCV and oxidative stress: implications for HCV life cycle and HCV-associated pathogenesis. Oxidative Medicine and Cellular Longevity 2016;2016:1–13. DOI: 10.1155/2016/9012580

[37] Olagnier D, Amatore D, Castiello L, Ferrari M, Palermo E, Diamond MS, et al. Dengue Virus Immunopathogenesis: Lessons Applicable to the Emergence of Zika Virus. Journal of Molecular Biology. 2016;428(17):3429–3448. DOI: 10.1016/j.jmb.2016.04.024

[38] Foppoli C, De Marco F, Cini C, Perluigi M. Redox control of viral carcinogenesis: the human papillomavirus paradigm. Biochimica et Biophysica Acta (BBA) – General Subjects. 2015;1850(8):1622–1632. DOI: 10.1016/j.bbagen.2014.12.016.

[39] De Marco F. Oxidative stress and HPV carcinogenesis. Viruses. 2013;5(2):708–731. DOI: 10.3390/v5020708.

[40] Gorrini C, Harris IS, Mak TW. Modulation of oxidative stress as an anticancer strategy. Nature Reviews. Drug Discovery. 2013;12(12):931–947. DOI: 10.1038/nrd4002.

[41] Cairns RA, Harris IS, Mak TW. Regulation of cancer cell metabolism. Nature Reviews. Cancer. 2011;11(2):85–95. DOI: 10.1038/nrc2981.

[42] Gupta SC, Pandey MK, Tyagi AK, Deb L, Prasad S, Deb L. Oxidative stress and cancer: advances and challenges. Oxidative Medicine and Cellular Longevity. 2016.; 2016:1 page. DOI: 10.1155/2016/5010423.

[43] Sosa V, Moline T, Somoza R, Paciucci R, Kondoh H, LLeonart ME. Oxidative stress and cancer: an overview. Ageing Research Reviews. 2013;12(1):376–390. DOI: 10.1016/j.arr. 2012.10.004.

[44] Dixon D, Edwards R. Glutathione transferases. In: The Arabidopsis Book. The American Society of Plant Biologists; 2010. p. e0131. DOI: 10.1199/tab.0131.

[45] Goto S, Kawakatsu M, Izumi S, Urata Y, Kageyama K, Ihara Y, et al. Glutathione S-transferase π localizes in mitochondria and protects against oxidative stress. Free Radical Biology and Medicine. 2009;46(10):1392–1403. DOI: 10.1016/j.freeradbiomed. 2009.02.025.

[46] Chen C, Wang DW. Cytochrome P450-CYP2 family-epoxygenase role in inflammation and cancer. Advances in Pharmacology. 2015;74: 193–221. DOI: 10.1016/bs.apha. 2015.04.005.

[47] Thimmulappa RK, Lee H, Rangasamy T, Reddy SP, Yamamoto M, Kensler TW, et al. Nrf2 is a critical regulator of the innate immune response and survival during experimental sepsis. The Journal of Clinical Investigation. 2006;116(4):984–995. DOI: 10.1172/ JCI25790.984.

[48] Kolls JK. Oxidative stress in sepsis: a redox redux. The Journal of Clinical Investigation. 2006;116(4):860–863. DOI: 10.1172/JCI28111.